THE PSYCHOLOGY AND BEHAVIOUR OF ANIMALS IN ZOOS AND CIRCUSES

By
Dr. H. HEDIGER

Director of the Zoological Gardens of Zurich,
Titular Professor in Animal Psychology and Biology of
Zoological Gardens in the University of Zurich

Translated by Geoffrey Sircom, B.A.

DOVER PUBLICATIONS, INC., NEW YORK

Published in Canada by General Publishing Company, Ltd., 30 Lesmill Road, Don Mills, Toronto, Ontario.
Published in the United Kingdom by Constable and Company, Ltd., 10 Orange Street, London WC 2.

This Dover edition, first published in 1968, is an unabridged republication of the Sircom translation of *Skizzen zu einer Tierpsychologie im Zoo und im Zirkus*, originally published in 1955 by Butterworths Scientific Publications under the title *Studies of the Psychology and Behaviour of Captive Animals in Zoos and Circuses*.

QL
785
. H366
1968

Standard Book Number: 486-62218-5
Library of Congress Catalog Card Number: 68-55533

Manufactured in the United States of America
Dover Publications, Inc.
180 Varick Street
New York, N. Y. 10014

Dedicated in respect and gratitude
to the perfect Zoo Director
MRS. BELLE J. BENCHLEY
San Diego, California

FOREWORD

Zoo directors seldom have the opportunity of doing much comprehensive research, occupied as they are by daily routine tasks. Occasionally, however, given the mind and the eye for it, they have the opportunity of recording chance observations, of rearranging these and of correlating them. Other zoologists, not in daily contact with animals from all over the world, have fewer chances of collecting factual material of this kind.

For this reason it may perhaps be appropriate to publish some more of these casual finds from the zoo, fully aware that they are merely fragments —scattered and unpretentious tesserae. To those from the zoo have been added others from the circus, together with still more from wild life. In addition there are ideas which occurred in the lecture theatre or during demonstrations before students on the living animal.

Many suggestions have been borrowed from the technical literature, notably from those scattered writings which may, from among many separate subjects, be classified as marginal or miscellaneous zoo territory. Most of these cannot be detailed, but only referred to. That is why I have tried to include comprehensive bibliographical references. These are intended to help the zoo or circus lover to enter more fully into the unique world of the animal's mind, should these notes inspire him so to do. They may well achieve this, since I am convinced that many surprises still await us in the field of animal psychological investigation.

To me, the animal psychologist seems like a cave explorer, who making his way through impressive tunnels, finds himself groping at the threshold of some lofty cavern, access to which will some day be granted to his astonished gaze.

CONTENTS

AN APPROACH TO ANIMAL PSYCHOLOGY

THE SEPARATION of animal psychology from human psychology, once regarded as normal, is becoming more and more exceptional. Instead we nowadays hear far more about comparative psychology, or about investigation into comparative behaviour, both of which subjects include the behaviour of animals as well as of man. The removal of this artificial, out-of-date barrier is proceeding rapidly, especially in zoos, the points of similarity of behaviour on both sides of the railings being too obvious to miss. How much do we see in the animal that is human—all too human— and how often do we regard the actions of *homo sapiens* as animal-like !

In the English-speaking world, the term comparative psychology, including men and animals, has long been current. From this point of view, animal psychology appears as just a branch of human psychology, in short, of psychology itself. If I am not mistaken, it was Goethe's friend, Carl Gustav Carus (1789–1869) who first spoke of comparative psychology in this sense, He even published a book with this title in 1866.

Every scientist, as well as every doctor, is familiar with the term comparative anatomy. It is not for nothing that in Switzerland every medical student begins his studies with comparative anatomy, both in the lecture theatre and in the dissecting room. It forms as it were the solid core of his knowledge of anatomy. The human organism is of unbelievable complexity. Is it not logical and appropriate that, in order to understand it better, the young doctor should learn about simpler organisms, the structure of which gives him an understanding of the elements, and a perspective of their increasing complication? Disturbances or malformations in the human body can often be understood because they may be compared with similar conditions in simpler organisms.

In addition to comparative anatomy and, in the broader sense, to comparative morphology, which are concerned with the internal and external structure of the animal, comparative physiology has long been in existence and is devoted to comparative observation of the functions of the organs, the organic systems, and the whole organism. The study of unhealthy organs (pathology), and many others, have long made use of the method of comparative observation. In this light, comparative psychology appears an urgent necessity. No one will deny nowadays that a knowledge of the psychological structure of simple organisms may be of use, being in fact indispensable for the fullest understanding of that most complex of creatures—man.

Yet it would be rash to assume that the only function of animal psychology is to provide a fuller understanding of man, that is, of anthropology. True, this is an important and sublime function of animal psychology, but not its only, or even its most important one. It is obvious that the animal psychologist must be in very close relationship not only to human psychology but in particular to zoology. If, at a pinch, animal psychology were conceivable without human psychology, it is absolutely unthinkable without zoology, which forms its indispensable basis. Whenever this fundamental fact has been ignored, it has led to gross errors of judgment, ridiculous exaggeration, or on the other hand to a serious underestimate of the animal's mind.

Animal psychology is firmly based upon zoology, not on theoretical grounds, but from logic and necessity. Starting as long ago as the time of Linnaeus (1707–1778) the famous Swedish naturalist and inventor of scientific nomenclature for zoology and botany, zoology consisted essentially of the minutest possible description of the various species of animals. Even today the exact classification of each species of animal—systematics—forms the indispensable basis for all zoological research ; for example, both the student of heredity and the physiologist have to know what species of animal they are dealing with.

The description of internal organs is naturally closely connected with the description of external characteristics. Investigation of the geographical distribution of each species followed the study of this kind of morphology. Animal geography came into existence, and soon not only the form and the distribution were being investigated, but also the efficiency of the animal and its organs. Physiology, or the study of the animal's metabolism, became indispensable. Similarly, many other branches of research developed, such as ecology, or the study of the relationship between an animal and its surroundings ; parasitology, genetics, etc. It was inevitable that not only should an animal's appearance, external and internal structure, distribution, and metabolism be studied, but eventually its behaviour as well. This inevitably led to research into behaviour, and this subject is basically confined to one aim ; the objective description of animal behaviour, or ethology. Some investigators, however, carried objectivity to excess, and imagined that they saw merely a resemblance to the working of a machine in the individual movements and reactions of an animal. Thus they overlooked many characteristics which animals share with mankind, rather than with machines.

Those behaviour students who regard the animal not only as a thing, an object as it were, but also as a sentient and acting being, a subject whose behaviour can be understood more or less personally, just as one man understands another, these are the true animal psychologists. Animal psychology may thus be described as the investigation of behaviour, plus sympathetic understanding.

Naturally there is considerable danger in this desire for understanding, yet given a critical approach it need not be excessive, since it is possible to keep a check, e.g. as in the making of prognoses, with their subsequent verification. If an animal psychologist thinks he can understand an animal, he must be in the position of being able to forecast its behaviour accurately in any given situation. The factual truth of this prognostication may be objectively verified. In the daily routine work of the zoo one is often able, as an animal psychologist, both to predict accurately the behaviour of an animal in a particular situation, and to take appropriate practical measures to forestall the kind of behaviour expected ; e.g. an inevitable fear reaction, or an aggressive attack. If the measures taken prove ineffective, the wrong prognosis has evidently been made.

Experienced animal psychologists and zoo managers with plenty of sympathetic faculties, or the right feel for the job—one might almost say the necessary biological intuition—are thus in the position to foretell certain types of behaviour which would take the outsider by surprise. For instance, it is possible to predict the path of flight of an animal, and the places in which it will hide. To do this one must put oneself in the position of the animal concerned, on the basis of exact ecological data, and ask what one would do in a particular situation if one were that animal. This may result in a surprisingly accurate forecast or prognosis. By this means alone, it has, for example, been possible time and time again to manoeuvre super-humanly strong anthropoid apes into a particular spot by means of a carefully controlled and graduated shock, such as by letting them see a small tortoise or snake.

In connection with the sympathetic approach to the prognosis of behaviour, animal psychologists have much in common with criminal psychologists. In crime, as in war, men revert in given circumstances to primitive conditions, often mistakenly described as animal-like ! The only animal-like thing about them is superficial behaviour, such as camouflage hiding and crawling on all fours. Motives and principles, however, are not animal but inhuman, or even human—a nice distinction !

The verification of the accuracy of prognosis of animal behaviour often corresponds to checking over problems in mathematics. There is certainly no lack of methods of testing ; they are only a little more difficult to apply than to problems on paper. If for instance an animal psychologist understands an animal well enough, and knows its usual reactions accurately enough, to predict with certainty a definite line of behaviour in any given situation, it is still by no means sure that the reactions predicted will then follow when there are a dozen critical observers waiting and watching. Indeed it is quite likely—highly probable in fact—that the presence of strangers will influence the normal behaviour of the animal under observation. The animal does not then react to a familiar situation, but to the disturbing presence of unfamiliar spectators. Anyone accustomed to demonstrating psychological facts about animals, either in zoos, or at congresses, knows how typical this result is. Hence the formulation of the theory that a psychological animal experiment cannot fail basically, but may often take a turn in an unexpected direction. The prudent animal psychologist therefore turns to good account a situation which has been changed by the presence of spectators. Apart from this there are other proofs than the evidence of trained observers ; in particular, photography and cinematography. These instrumental aids are of capital importance to animal psychology. Just as the histologist verifies his observations by a large number of prepared sections of tissues, the animal psychologist too must be able to follow up his observations with objective pictorial documentation. Yet even in the case of histological preparations, genuine and unfalsified, lying on the stage of the microscope, there may be differences of opinion in interpreting what is visible in the enlargement.

If the possibility of subjective understanding between two human beings can exist—nowadays hardly a matter of doubt—it must also exist under certain circumstances between man and animal. There are plenty of arguments to support this, but we confine ourselves to mentioning a few that are specially appropriate. Let us, for sake of illustration, take the case of a good animal tamer putting on a difficult act with trained tigers. This man's efforts would be inconceivable unless he had remarkably deep sympathy for each of his animals. As I hope to show later on, he must keep an extremely watchful eye not only on the character and individuality of each animal, but also on its mood of the moment, and on the delicate and complicated interplay of relationships between animals, and of every animal to all the others. He must calculate to a nicety the result of a change in the lighting, a sudden unexpected movement on the part of one of the audience, noises at close range, and so on, and above all, must watch each animal's expression most carefully. In short, the trainer must thoroughly understand an animal in every situation in order to be able to work with it successfully. This may at first seem improbable or impossible to the outsider. Who can tell the meaning of a gleam in a tiger's eye, or what its snarling signifies ? Through continued intimacy, constant observation, and countlessly calculating and testing, a man with a flair for this sort of thing may in fact acquire a really thorough understanding of his four-legged partner. Naturally, nobody can be expected to understand an animal at first glance, and even the experienced can sometimes make mistakes. Yet are there any experts in human psychology who consider themselves infallible in their own particular line ? On many trips among primitive races, that is, natives who had scarcely ever been in contact with white men, I soon realized that to understand them without the help of a linguistic medium was a similar problem to that of understanding an animal.

Somewhere in the heart of New Guinea were men who, on seeing a white man, fled into the bush screaming with fear, thus showing a true flight reaction exactly like a shy wild animal's. Others again could not eat in the presence of strangers, a familiar feature with many animals. To some, a piece of wood with a hole in it represented a personification full of significance. It was not just a piece of wood, but a representation of their ancestors, or of a demon. On the Sepik River, we came across a gigantic crocodile made of palm branches and grass in which some adult men were concealed. The crocodile devoured, in anguished ceremonial, the whole of the young men of the tribe. Again, there were monoliths which had never been touched by any natives because they were afraid that if they did so, they would fall down dead. Is it not obvious that the ethnologist must not only describe the behaviour of his natives, but understand them as well ? He can interpret the initiation ceremony of the young men, or see the importance of a twirling stick or a monolith, even though he can hardly make himself understood by the natives, and then only indirectly.

The lack of direct contact in a speech is not a fatal obstacle to a sympathetic subjective understanding of behaviour in man or animal. There is no need to go to natives in foreign lands—we only need to consider our own babies and tiny children. We cannot cross-question them, and yet we can to some extent understand them.

This short digression is just to show that an understanding of creatures different from ourselves is possible. Such an understanding is at best only partial, yet we should lose something vital if we failed to attempt it. It would be both inexcusable and unscientific to neglect on principle such an opportunity.

One of the greatest dangers in trying to understand an animal, our tendency to humanize, must now be dealt with ; and here I would like to mention another difficulty ; there is only one species of man—*homo sapiens*—but about a million different species of living creatures, with a few hundred more being discovered every year. This means a million different kinds of behaviour, and a million different psychological studies ! It will be a long time before this gigantic task has been appreciably mastered.

When an animal psychologist has for example made some interesting discovery about a mammal or bird, it will probably happen that his observation will be the subject of generalization. To the animal psychologist, trained in zoology and aware of the vast number of species, this danger is less than to the psychologist who approaches it from the standpoint of human psychology, and who fails to realize that there are thousands of different species of mammals, and thousands more of birds. And then they are apt to say, airily : mammals behave like this—or birds like that. Up to the present, very few species of mammals or birds have had even a cursory psychological investigation. The mammals most closely examined are the white rat, the chimpanzee, and the dog, while among birds the best known are, of course, the hen, some gulls, ducks, and pigeons. The psychology of many thousands of others is either not known at all, or only fragmentarily.

That is why I consider it an extremely important function of animal psychology to discover as many examples as possible of genuine and universal conformity to laws and rules of behaviour, valid for complete groups as well as for individual species. While certain deserving individual investigators are working on an accurate inventory of the behaviour of one or more species of animals and investigating their psychology as minutely as possible, others, less specialized, are attempting first and foremost a psychological study of whole groups. This kind of research is most appropriate in zoos, since universally valid rules of conduct appear here far more valuable to the observer than exhaustive research into one particular species.

All this is not by any means intended to give the impression that this line of research—animal psychology in the zoo—claims to be the sole means of grace. There are of course other very different and most valuable introductions to animal psychology, and above all different methods used in animal psychology, *e.g.* comparative psychology. Only one has been chosen here, and that one deliberately. Thus it is necessarily one-sided, and is in fact stressed as such, since after all, this book is about animal psychology in the zoo.

Here I should like to summarize some of the most suitable of the many possible introductions to animal psychology.

(1) One may for instance choose as a starting point that apparently humblest and simplest of all animals, the protozoon, the primitive, unicellular animalcule ; then, as we gradually ascend through the animal

kingdom, we can consider step by step the gradual additions to and complexities of the psychological phenomena ; in other words, the whole animal kingdom from amoeba to gorilla. This method was described in 1926 by Friedrich Hempelmann in his animal psychology, which ran to 676 pages. No doubt this procedure is perfectly logical ; but one must realize that the minutest microscopical animals are most remote from us. In their case, the possibility of understanding is least because their resemblances to man are minimal. We gain an insufficiently clear view of the life processes of these tiny creatures through the clumsy barriers of the microscope. How can we gain a sympathetic understanding of a creature the size of a speck of dust, almost invisible to the naked eye ? Yet this was attempted in 1906 by a pioneer of animal psychology, H. S. Jennings, Professor of Zoology at Pennsylvania University, in what might be called a charming way.

Jennings, who for many years was engaged in an intensive study of a minute infusorian, the paramecium, ventured to describe its daily life in one chapter of his comprehensive and greatly appreciated work *The Behavior of Lower Organisms,* on the strength of his intimate knowledge of this microscopic creature. For example, in his combined apologia, based upon many factual examples, he describes how one of his favourite infusoria was nibbling at a mass of bacteria (zoogloea) in a small pool of water that was lit by a ray of sunshine.

" The water gets warmer and warmer, and in a short time our infusorian starts to move a little, turning round and changing position, yet always keeping close to the Zoogloea. All the free-swimming paramecia have long since left this area. As the water gets warmer, our creature suddenly leaves the mass of Zoogloea and dashes madly about under the influence of the great heat. First it swims backwards, then forwards, trying one direction after another. By luck, one of these directions leads it into a cooler area. It keeps on swimming in this direction and its behaviour becomes more settled ; now it swims around quite quietly, as at first, until it finds another cluster of bacteria, when it starts feeding all over again."

Jennings then continues by asserting that the paramecium's behaviour was virtually the same as that of a deaf blind man, or someone groping in the dark. This comparison, however, is not valid ; the paramecium has none of man's conception of space ; it cannot predict, gauge or calculate. In his eagerness, perhaps oo in his passion for his pet subject, the scientist has considerably overshot his mark. He has clearly not avoided the pitfalls of sympathy, and of a determination to understand, so often the case in former centuries.

An important part is played here by absolute size ; the bodily dimensions of the animal being studied. The smaller it is, the greater the difficulties of observation and interpretation. The significance of the size factor will be realized at once if we think about Tom Thumb—that fairy-tale figure with every human attribute, but as tiny as a thumb, or even tinier. What difficulties arise when we try to understand this little creature ! His features, gestures, voice, complexion, breathing, all the finer shades of his behaviour and expression, suddenly become largely inexplicable, even imperceptible to us. Is the little dwarf laughing or crying ? Is he asleep, dead or unwell ? Is he joking or in deadly earnest ? Here the small size of the object, and that alone makes observation and understanding extremely difficult for us. That is why the protozoa, the smallest of all animals, are so infinitely remote from us, quite apart from their fundamentally different body structure. Thus an introduction to animal psychology that starts with the simple animalculae presents certain difficulties. These are well illustrated in the classic situation to which the researches of F. Bramstedt (1935) and U. Grabowski (1939) into the behaviour of the paramecium have led. Both authorities, representing two different schools of animal psychology, tried to find out the ability to learn of this fascinating infusorian. Each came to a completely different conclusion. Here too perhaps, although only unicellular organisms are concerned, the sympathetic attitude of the research worker towards the animal studied, plays a part, as it occasionally does with decisive importance in studies on higher animals.

Bramstedt conceived a delightful experiment—a kind of " micro-circus " as it were. On the cover of a double glass container the same size as a slide (76 × 26 mm), he put a drop of water so that one half of the drop remained at a temperature of 15°C, thanks to the mass of water under-neath being kept constant at that temperature, while the other half was kept at 42°C by a similar arrangement. The temperature in either half of the water drop could be measured accurately by thermo-couples, and in addition the drop could be darkened or illuminated at will. A system of mirrors permitted accurate plotting of the swimming tracks, and a computing stop-watch recorded the times taken in each half of the drop.

Thus, when the warm half was illuminated and the cool half kept dark, a paramecium in this miniature circus-ring only needed between one and one and a half hour's training to learn finally to avoid the half that was uncomfortably warm for it. Then, after the training was complete, or when the animal was fully trained, both half drops were kept at 15°C, and the half drop that had previously been hot (uncomfortable) continued to be avoided by the paramecium. This can only be explained, says Bramstedt, " by the fact that a light-heat association had been formed."

What had been learnt lasted at any rate for only 15 minutes or, to use the terminology we shall introduce later on, the state of training continued for only 15 minutes. After that, the infusorian had forgotten once more what the boundary between light and dark meant. Bramstedt's paramecia, however, learnt even more than that ; for instance, they learnt to recognize the shape of a container of triangular or rectangular section, provided that it was not too large. If it was limited to a circle of 10 mm diameter, it could still be " intracentrally conceived as a unit " by the infusorian. Thus if a paramecium, for example, swam about for two or three hours in a tri-angular container, and was then quickly transferred by means of a small pipette to a circular one, it still continued to swim along a triangular course for some time. The same sort of thing occurred on transferrence to containers of different shapes. Bramstedt adds, " Paramecia therefore have the faculty of learning the shape of space."

Grabowski, who checked Bramstedt's astonishing discoveries of the learning powers of unicellular organisms, came to completely different

7

conclusions, so it is not difficult to compare their main points :

Bramstedt	*Grabowski*
1. Training means success in learning.	1. Success in learning is only simulated.
2. Circular, triangular and rectangular spaces are differentiated.	2. There is no evidence of any such capacity.

The question of educability is very clearly linked with that of a central nerve organ ; yet the answers to it are conflicting, not only in the case of protozoa, but also with radially symmetrical metazoa, *e.g.* starfish. A very simple observation in the zoo shows, however, that certain starfish (*Asterias*) must have a central organ, since the animal reacts not only as the sum of its five arms, but as a centrally motivated unit as well. In the small aquarium at Basle Zoo, we had under observation for some months a starfish that was active only at night, when it used to crawl round the tank. By day, or when surprised by switching on the light at night, it would always retire to one particular spot under a projecting stone slab. This spot could therefore be correctly called its home, and I have frequently demonstrated this attractive effect to my students.

Returning regularly to one special spot in the area, under widely different conditions, can only be explained on the basis of central motivation of the starfish. Thus, simple observation of the starfish in the zoo aquarium compels one to assume a central nerve organ. Regular return to the home month after month can never be the result of five arms reacting independently.

(2) After this excursion into the realm of protozoa and radially symmetrical animals, let us examine another possible introduction to animal psychology. In the first few years of this century, when the newly founded science of animal psychology was enjoying its greatest successes, there was one particular piece of research apparatus in use in laboratories which long ruled supreme, and which was expected to give the key to the animal's mind. This apparatus, considered to be practically all-powerful, was the maze.

It seemed in those days to fulfil the fondest hopes of the animal psychologists, and consisted of a universal piece of apparatus in which any sort of animal —fish, bird or monkey—could be harnessed, so to speak, and its intelligence and mental characteristics read off in figures at will from a kind of manometer, after certain adjustments had been made. This idealized picture is perhaps rather exaggerated, but at any rate it gives some idea of the attempts of scientists to find a uniform experimental set-up for all animals, to give directly comparable results expressed, if possible, in decimals.

Strangely enough, it did seem at one time as if this illusory idea, now considered rather naive, could really be made to work by means of the maze, developed from an idea of the American psychologist W. S. Small. This idea, first put forward in 1901, consisted of observing the behaviour of various animals in mazes suitable to their size, using the plan of the famous Hampton Court maze as a model.

Small first used this sort of maze with rats. The paths were just wide enough to allow a rat to pass along comfortably. This experiment was considered suitable for rats, and therefore biologically sound, since wild rats are used to running along branching passages and remembering their way. The problem put to the four-legged candidate consisted of getting from the outer entrance, where the rat was set down, and reaching the centre as fast as possible, where a tempting bait waited as a reward. At first the animal naturally hesitated, ran along blind alleys, and so on. But things improved with almost every repetition, and eventually an experienced rat would run at incredible speed, avoiding every blind alley, and make straight for the goal, where it enjoyed the titbit, or regained its nest.

According to the maze-psychologists, objective measurements could be taken both of how long the rat needed for each run, and how many repeats and mistakes were necessary before the shortest route had been learnt. Small's discovery had great success, especially in America. The maze was regarded as a normal piece of apparatus in all experiments in animal psychology, and was thereafter modified time and time again. Tall mazes, deep mazes, right-handed, left-handed, circular, multiple and T-shaped. *etc.*, were invented.

Famous institutions specialized exclusively in maze experiments, and vast numbers of reports were published on the behaviour of animals in all sorts of mazes. " Countless thousands of living creatures from babies to toads and cockroaches have, thanks to Small's experiments, been made to run, waddle, or swim through countless mazes. The victims have been subjected to hunger and thirst, have had nerves severed, and sensory organs removed ; they have been injected with poisons, deprived of vitamins ; have had their memories tested, as well as their power of transferring to other mazes what they have learnt, *etc.*" Thus the Dutch animal psychologist J. A. Bierens de Haan (1937, p. 2), who published an admirable summary of experiments with mazes.

However, the maze method must be regarded as superseded nowadays. It failed, naturally, to fulfil the exaggerated hopes originally put in it. Serious doubts have been raised as to whether the enormous number of experiments made was commensurate with their value. At all events, in 1933 Norman L. Munn of Pittsburgh University chose the maze, in a comprehensive account, as an introduction to animal psychology, *i.e.* the behaviour of the white rat in the maze. The white rat is by far the most thoroughly investigated animal in experimental psychology. This very fact allows us to study more closely another possible introduction to animal psychology.

Before we do so however, it must be emphasized that the maze method has been mentioned here merely as an especially representative method of a particular trend of experimental animal psychology, restricted to the laboratory. Similarly, various experimenters hoped that the discrimination apparatus, the problem box, and the delayed reaction would, like the maze, have a wide field of application within the systematic groups, and would give directly comparable results, permitting equivalent data on memory, intelligence, *etc.* for the most widely differing species.

W. Fischel (1938) made an admirable summary of such methods, showing, on the basis of Hunter's experiments, how much the performances of primates under delayed reaction improved, from the prosimians through the New and Old World monkeys to the anthropoid apes, just as we might expect (or rather might previously have expected) from the systematic positions and the brain differentiation of the groups. Nowadays we know that the zoological and psychological systems of animals cannot be considered identical.

The value of the method of delayed reaction has therefore undergone considerable modification recently, all the more since it has become apparent that the period of delay in certain insects is greater than in the anthropoid ape (Orang outan), as N. Tinbergen (1951) says.

I myself performed delayed reaction experiments in 1935 on an African elephant (v. Fig. 1). A titbit was put into one of three identical covered pails in sight of the experimental animal, watching from its stall. Then its door was closed for a while so that the elephant could not see the pails. Next, the three pails were exposed and the animal could claim its reward provided it had managed to remember the right pail, go up to it and open its lid. Thus was established—at any rate theoretically—the period of delay of the animal ; in other words, how long it could remember the right pail.

In practice, however, many difficulties arise. For instance, the animal may develop a unilateral preference before it understands the situation, e.g. it may keep on choosing the left-hand pail once it finds the reward there. Many experimental animals, as in the case of my African elephant, soon lose interest and get excited under compulsion, their actions thus becoming valueless.

Serious sources of error also occur sometimes through the experimenter involuntarily giving the animal hints (cf. Chapter 11), or through it getting used to a particular sequence in the choice of pail. The first risk may be met by not letting the experimenter know in which pail an assistant has put the reward ; the second by adopting an irregular choice of the positive pail, according to a scheme drawn up before the start of the experiment.

As with various other experiments from which a profound insight into animal psychology was hoped, the delayed action experiments proved to be too stereotyped, as well as not equally practicable for all groups of animals. Mental capacity, as we have said, is not distributed throughout the animal kingdom according to systematic position, as was once rather naively believed, but much rather according to the particular life habits and environment of a species of animal. Thus for example, the peacock is far superior to the dog as far as making a detour is concerned, i.e. in indirectly approaching its goal. The chicks cannot always follow their mother directly through the branches. The sand wasp is far more tolerant of delay than the ape or the elephant. The insect, when caring for its young, depends upon this tolerance, as Baerende (1941) has shown in detail.

(3) It is understandable that those animals with which man comes into closest daily contact at home, on the farm, domestic animals, should be chosen as a starting point for animal psychology ; such animals as the cat or dog, horses, cattle, sheep, goats, rabbits, guinea pigs, etc. We must not forget, however, that there is a complication here : none of these animals are pure natural products, so to speak ; thanks to the results of centuries

of breeding by man, they have undergone definite changes in the course of domestication. Therefore domestic animals are artificial material for a starting point, not a natural one. Their modes of behaviour must not, without further consideration, be applied to the hundreds of thousands of undomestic animals that form by far the majority of creatures living today, and which provide the ideal research for us, as they are the only genuine primary material.

As we have seen in the case of the white rat—a domestic animal, and the favourite animal for maze experiments—a domesticated animal may be fundamentally different from its ancestral wild stock. Since the wild rat was much more difficult to handle, it was replaced in 1895 by the far more docile standardized white rat in laboratory experiments in animal psychology. Thus, instead of making the experimental method fit the animal, the fatal mistake of breeding an unnatural domestic, or laboratory animal, a sort of abstract animal form, nicely adapted to the experimental apparatus, was made.

Handling domestic animals is always easier in practice than handling wild ones, but we must beware of thinking that we can study animal psychology from domestic animals alone. Natural, genuine behaviour, all-important for our purpose, can only be studied in wild animals.

(4) Another possible introduction to animal psychology is to begin with its historical origins and to finish up with the results of modern experiments. This is certainly a most attractive method, and was followed for example by Heinrich Ernst Ziegler in 1920. But as animal psychology is making somewhat rapid progress, the historical view, fascinating as it is, may lead to the danger that present-day problems get scant attention. The basic matter of animal psychology may be reached too hastily, and too late.

(5) There is a fifth possibility, an attractive and fundamental problem, moreover ; that of choosing as starting point a really basic question of animal psychology, the problem of instinct. Countless investigators have begun on this basis, *e.g.* J. A. Bierens de Haan in his book of 478 pages, published in 1940, *Animal Instincts and their Modification Through Experience. An Introduction to General Animal Psychology.*

The word instinct raises not only a central problem of animal psychology, but is one of the most hotly disputed. It is thus open to doubt as to whether a topic on which so much dissension is rife makes a good introduction to this subject. The plasticity of those instincts referred to in the title of this book has already given rise to lively discussions, since scientists such as Konrad Lorenz (1937) are vehement in their support of the opposition view, namely, that instincts, or instinctive actions, cannot be modified at all ; that they are not even plastic, but completely rigid.

Many think that instinct, or instinctive action, approaches the miraculous. For instance, how does a baby opossum, born deaf, and in all respects embryonic, find its way so unerringly into its mother's pouch ? For my part, I find the pouch itself something of a marvel, as well as the fact that an opossum, exactly similar to both its parents, should be produced from two microscopic opossum cells, one male and one female. Not only the animal's behaviour, but also its structure and origin, fill many observers with wonder, causing them to ask "Why then should those inborn directions for use that we call instincts be more marvellous than the creature in which they are innate ? "

But almost every investigator has his own views and definitions of the concept of instinct ; such great confusion reigns that well-known scientists such as David Lack (1946, p. 189) have decided to give up the use of the word altogether as being needlessly and even dangerously confusing and misleading. Carl G. Hartmann (1952, p. 99) has come to the same conclusion. For this reason, any concept as debatable as this is of limited value as a point of departure for the study of animal psychology.

N. Tinbergen's excellent book on *The Study of Instinct* (Oxford, 1951) did much to clarify and re-orientate the subject. There is no doubt that Tinbergen, together with Lorenz (1943) stands in the front rank of those who are working on instinct. Tinbergen's work is characterized by a highly objective analytical approach. Zoologists, physiologists, psychologists, anthropologists and philosophers owe him a great debt for this summary of his astonishing research, especially on sticklebacks and seagulls, as well as for having incorporated in his material the latest researches of his fellow workers at present available on the subject. Thus his book gives us an indispensable cross-section of the present state of our knowledge on the complex theme of instinct.

According to Tinbergen the objective investigation of behaviour, or ethology, merges in the last analysis into neurophysiology. This line of investigation is without doubt completely justified, and is a most fruitful one. However, in the zoo it is obviously rather remote from practical needs and potential uses.

(6) In view of these facts, it would seem not entirely out of place to blaze a trail of my own in the field of animal psychology ; that, in fact, of the animal psychologist in the zoo. He, perhaps, has special qualifications, enabling him to supplement and compensate for others, who may also be one-sided. At any rate, no more natural elementary basic material exists than the wild animal ; by which we mean not just predatory animals, but all those animals which have developed in nature without man's interference. The wild animals most suitable for our purpose are the ones that have been least influenced, or least distorted, and are thus most genuine. They form the foundation, the measure, the norm of all our estimates of animal behaviour.

Naturally this is true first and foremost for wild animals living in freedom, not for those in captivity. Nevertheless, to the zoo animal psychologist, the free wild animal's pattern of life forms a subject of the greatest interest. Not only does he obtain many of his animals from the wild, but gets hints on their care as well. Capture frequently takes place before the animal is removed to the zoo, and this, in turn, is often preceded by observation in the wild. So there is a clear justification for attempting an introduction to animal psychology from this specific viewpoint. A warning must be given at this point, however ; here too, we are concerned with a one-sided and incomplete form of animal psychology—that from inside the zoo ; and this sketch is no more than an outline of it.

2

HOW ANIMALS LIVE

THERE was a time in the history of animal psychology—a time that extended into our own century—when this science consisted to a large extent merely of unverifiable anecdotes, many of them based on illusion, but all of which were highly anthropomorphic.

By way of illustrating this anecdotal phase I should like to quote a passage from Peter Scheitlin's two-volumed *The Study of the Souls of Animals* of 1840. This book, significant for its own day, was, incidentally, published by Cotta's, the firm that printed many of Goethe's works. On page 175, for example, the author says about the elephant : " The elephant by nature is peculiarly affectionate, attached and trustful, and its soul is without guile. How friendly it is towards the horse, whose nobility it acknowledges ! How even tempered with its keeper ! And how dear are children to it ! It is truly concerned for them. Yet such love, such care for children we have observed only among female elephants, that are gentler and of better nature than the males. This tenderness of theirs renders them far more sensitive to music also. The elephant is passionately fond of it. Many tunes are known to affect them deeply, both males and females, exciting them to sympathy and love in a way unknown to any other animal."

Naturally, this sentimental, over-humanized kind of animal psychology led to forthright criticism in the second half of the nineteenth century. In 1894 Lloyd Morgan, the representative British animal psychologist of his day, came out strongly against it. Through his influence, contemporary investigators formulated a basic law, known among animal psychologists as Morgan's Law, which became universally accepted. This law states that an animal's actions must never be considered as a higher psychical performance, but must always be ascribed where possible to simpler elementary causes. Thus for instance the functionally perfect hexagonal shape of the honeycomb should not be attributed to the bee's mathematical calculations. Instinctive actions must not be confused with intelligent performance.

Morgan's Canon, or the " Principle of Economy," as it has been called, has been overdone. Finally, even intelligent learnt actions based on individual experience have, through excessive reluctance to recognize genuine higher actions, been all too easily disposed of as automatically controlled actions, under such catchwords as reflex ; reflex chain ; tropism, *etc.* So great in fact did this over-compensation become, that cautious scientists like the Dutch Bierens de Haan (1940, p. 17) rightly pointed out the importance, in animal psychology, of explaining an animal's performance and behaviour as correctly, not as simply, as possible.

Morgan's Law, it is true, has done much to remove the study of animal behaviour from the realm of sentimental anthropomorphism, and place it on the solid ground of exact observation. We must state at once that in animal psychology the principle of economy in many cases is really justified, not only with regard to elucidation, but above all in regard to lines of enquiry. We shall not go out of our way here to ask remote questions, but make use of this principle of economy not only in our explanations, but in our enquiries as well.

One of the most straightforward and simple questions that an animal psychologist can ask is this : " Where does the animal really live ? " With regard to the larger species, we can usually say—away from towns ; where there are few or no people ; in fields and woods ; on grasslands and in the jungle ; in the ocean or in the polar regions. Animals and men are for the most part mutually exclusive. Many species, once common everywhere, such as the raven and deer, have retreated to the mountains. What is true for Central Europe is also unfortunately true for Java and Brazil, India and Central Africa. The world of the larger animals is gradually contracting everywhere, forcibly constricted by technological changes in the landscape. Nowadays they live mostly in regions to which they have withdrawn, or in reserves generously provided for them by man. Their original living space yields before artificial landscape. The animals most affected are naturally the largest ones, since the small ones, needing less space, are less affected.

This simple statement of fact at once implies that the presence of animals is now mostly secondarily, not primarily, conditioned. Even the animal in freedom is more or less influenced by man, if only in the sense of being restricted to within a retreat. Proper observation of an animal's life and behaviour can only succeed if these various animal–man relationships are allowed for. Thus, animal psychology is essentially the study of animal–man relationships. We can have little or no knowledge of pure original animal behaviour ; nearly everything is now permeated with the animal's reactions to mankind, and to his technical achievements. Nowadays, practically all animal behaviour is directly or indirectly influenced by man.

Direct influences include hunting and fishing. Anyone who looks up the Hudson's Bay Company's fur pelt records for the last few centuries, or elephant shooting statistics (10,000 to 20,000 annually in the Belgian Congo alone), or the records of whaling, or any other hunting records, will at once realize the ecological importance of man. In addition, there are the slaughter of Arctic seals, sea fishery, etc., all of them decisive factors.

Indirect influences include, for instance, changes in landscape, agriculture, the building of towns and roads, construction of reservoirs, woodland clearance, drainage, forestry, high-tension cables, dams, canals, changes of vegetation or amount of cover, the introduction of plants and animals, and so forth.

It is very illuminating to remember that in historical times there were still wild cats, lynx, beaver, bear, moose, bison, wolf and lammergeyer everywhere in Central Europe. In the heart of these animals' former territories lie cities, railway marshalling yards, and airports, with their sprawling blocks of buildings. The picture is little better in other continents. About a hundred years ago, near the site of Sydney Central Station, the emu, an Australian ostrich, was first discovered. In Africa, it is hard to find a single adult elephant that has not been shot at. All this implies a fundamental change in animal behaviour ; secondary behaviour, *i.e.* behaviour influenced by man, is now practically universal.

Generally speaking, it is the rule that where there are many people, there are few animals, and *vice versa*. Yet if only we keep our eyes open, we see that even in the middle of towns some wild animals survive—in parks, in gardens, even on or in our very houses. The house mouse ; the field mouse ; the house rat ; the brown rat ; the house martin ; the swallow ; the blackbird ; the sparrow ; the sea gull ; the stork ; the weasel ; the stoat, and so on. Several species approach the outskirts of towns, *e.g.* roedeer, badger, fox, rabbit, buzzard, *etc*.

Thus we can differentiate between two different ecological animal groups, which at the same time are fundamentally different psychologically (*see below*—Different psychotopes).

Shy of man's influence	Accepting man's influence
Technophobes	Technophiles
Majority	Minority

The first group is distinguished by a negative attitude to man's technical improvements, the second by a positive attitude, sometimes taking advantage of it when possible. Even within a single group of animals (*e.g. mustelidae*), one species may be a technophobe (pine marten), the other a definite technophile (the common marten, *Martes foina*). Here we encounter two completely different mental make-ups. Clearly we can form a better prognosis for technophile animals ; they are adaptable and know how to make best use of all sorts of artificial environments—houses, dams, canals, electric pylons, barbed wire, ditches, arable land, *etc*.

The technophobe animal—essentially the free wild animal—has basically two means of withdrawal from man's disturbing influences. It can change either its place or its time. In districts where they are greatly exposed to persecution, many diurnal animals become nocturnal, changing their times of activity. The other possibility is to give up the familiar space. This is no doubt the more drastic way of avoiding man with the least favourable prognoses.

This brief outline of the question of where an animal lives, with reference to man's influence on it, is followed by the second equally simple question—how ? Yet it is by no means as easy to answer as it might seem at first sight, for we know astonishingly little about the appropriate details of the animal's life.

It is important to consider the milieu—the surroundings in which it lives—not only in human psychology, where its importance is universally recognized, but in animal psychology as well. Domestic and laboratory animals do not live in their original milieu, but in a secondary man-made environment. Their behaviour is correspondingly distorted and unnatural.

We can see the need for insisting on this question of where and how an animal lives in the astonishing assertion—almost grotesque to a close observer of nature—made by a well-known student of human psychology, Pierre Janet (1935), when he said that the chief differentiation between animal and man is, that the former possesses no street system. Anyone who has observed animals, especially mammals, in freedom, knows how untrue this statement is. In actual fact, countless animals had their own street systems long before man existed, and in many cases man has adopted animals' tracks, that is, he has shared them, and gradually re-shaped them into human paths and traffic routes. To take an extreme case for example ; in North America many continental highways, and even railways, follow immemorial bison trails, as Martin Garretson has shown (1938, p. 54). In Africa to this day there are many rarely visited districts where animal tracks constitute the most convenient traffic routes for men—often the only practicable ones. I know from personal experience in Central Africa how pleasant it is, after stumbling painfully along through trackless bush, to strike all at once a hippo's track, running comfortably through the landscape like a well kept garden path.

But these streets, or animals' paths, of course are only lines of communication between single fixed points, the homes and colonies of the animals. Let us take a look, then, at the animal's home. It is surprising how much in common animal and human homes have, basically. In this similarity of subjective organization and of living space arrangements certainly lies one of the closest links between animals and man. In the animal's home, there are often special eating and drinking places, bathing and sleeping quarters, food stores and lavatory, sun-bathing terraces and nurseries, etc., etc. One locality only, the fireplace, found in even the most primitive man's hut, is absent in the case of every single animal.

One of the most decisive differences—perhaps indeed the most significant of all differences—between the animal's home and man's, lies in this fundamental distinction, seen from this particular viewpoint. The conquest of fire was literally the spark which originally kindled not only man's technique but his culture as well.

It may be worth while to summarize the latest research on the animal's home. A great deal of investigation was of course necessary in order to find out the natural living conditions of animals, and this work could not be carried out in the laboratory. Here the animal does not fit into the experimental arrangement, and the investigator has no choice but to follow up the animal, as quietly and as inconspicuously as possible. The prospects of answering the simple, but long-neglected questions about animals' living conditions are only possible if they are studied in undisturbed freedom, in their natural surroundings.

First and foremost, it is the field biologists who have done the best pioneer work here, in our own native woods and fields, as well as on the distant steppes, and in tropical rain forests. With the help of camera and binoculars, these men have chosen as subjects for exact observation individual animals, or else organized groups such as packs and herds, and have noted down in the minutest detail, with perseverance and devotion, the common daily life of a bird, a beast of prey, or a monkey.

With the objectivity of white-coated laboratory biologists taking readings from their precision instruments, field biologists on trek somewhere in the virgin forests of Indo-China, or the heart of Central America, have analysed the habits of gibbons, or the daily life of a group of howler monkeys.

For the outsider, perhaps one of the most surprising results of this field research is the fact—observed and confirmed countless times—that, with few exceptions, the so-called free-living animal does not wander about at

Figure 1a.—*African elephant in delayed reaction experiment for testing memory power.*

Figure 1b.—*The expected reward is in one of three similar containers. The elephant has just opened two of the tubs.*

Figure 2a.—In the zebras' territory, special tracks lead to the " termites' nest " used as a rubbing post. This is an important fixed point in freedom, as well as in the zoo.

Figure 2b.—A considerable part of the zebra's day is devoted to rubbing against a specially provided " termites' nest ", in the interests of body-care.

will, but clings with remarkable tenacity to its little plot of earth, to its home, and usually leaves it only under the strongest compulsion. Such a home, that piece of land possessed by an individual, or shared by an organized group of individuals—a pack—is called in biological terms a territory. It is moreover vigorously defended against other members of the same species, especially of the same sex, and far less so against animals of different species.

Thus the fox, for instance, usually tolerates the presence of a badger in its neighbourhood, but would snap at another dog fox, and chase it away. The eagle even allows small song birds inside its eyrie, where they may sometimes nest, but another eagle would be remorselessly driven off by the owner. The swan, whose territory is about one third of a square mile, allows coots, little grebes, and gulls on its terrain, but not under any circumstances other male swans. The squirrel does not mind hares, wood mice, or stags in its woodland home, but chases off in a rage all males of its own species that dare to set foot on its own territory.

In this way the territories of different species of animals may often overlap, or even coincide, though animals of the same species are forbidden the area. The territories of animals of the same species, for example foxes, blackbirds, or lions, would seem like the separate stones of a mosaic, if we could see them from the bird's eye view, each the same colour, and about the same size. Every mosaic of this sort has then superimposed on it several others, the territories of different species of animals, with stones of different colour and size. In the territory of a herd of chamois live marmots, jays, alpine jackdaws, and mountain salamanders as well. Different colours can overlap, but not stones of the same colour, as this would mean fighting, sometimes ruthless fatal combats.

Basically this is really rather like what happens among men. If a pigeon or a crow perches on the roof of our house, or a squirrel scampers round the garden, or a cat wanders in, we are usually indifferent. On the other hand, if we meet an uninvited member of our own species in the house or garden, our feelings of ownership are strongly aroused.

While small animals such as mice, lizards, or inch-long fish, have territories of up to ten square yards, lions or tigers need a dozen square miles or so. A couple of squirrels have a smaller territory than a couple of roe-deer, corresponding to their smaller food requirements, while a quail needs less than a hawk. Carnivores always have larger territories than their herbivorous prey. For example, there must be room for many mouse territories in the territory of one fox, and for many antelope or zebra territories in that of a lion, since these herbivores must keep the carnivores self-supporting, so to speak, if the balance is to be maintained. Thus the size of an animal's territory is mainly dependent on two factors, the food requirements of its inhabitants and the products of its soil, either in the shape of animals to prey on, or food plants.

The size of a territory is not determined by the animal's desire to own a nice big park. It is even demonstrable that the territories are usually smaller in districts where food is abundant, than in those where food is scarce. Here one might mention in passing that in the zoo, where the animal is entirely relieved of the need to seek out its food, this being fully provided for it every day, the living space, the artificial territory in other words, can be very much smaller still.

In addition to the problem of the amount of space needed by an animal, the problem of its quality has long been neglected (Hediger, 1950 ; pp. 71 *et seq.*). Once again we are indebted to the field biologists for the essential information. To the animal that lives in it, a territory may not be of equal value from one end to the other, but is in fact in various ways subdivided into different localities, each associated with definite functions and significance. As we have pointed out, there are sleeping quarters, bathing places, food stores, *etc.*; in short, special inside arrangements, differing for various species.

Knowledge of these conveniences, if I may use the term, is naturally of extreme importance in zoos, since one wants the inmates to feel as comfortable, as snug, and as much at home as possible. A small illustration may show what I mean. In Africa, I noticed that many termites' hills, standing up above the grass on the savannah, seemed to have their tops polished or worn away. It then transpired that elephants, buffaloes, and, above all zebras, of the neighbourhood used to come regularly and rub themselves luxuriously against these decorative termite castles. Often, round the bases of these cement-hard humps built up to six feet or more by the insects, great tufts of buffalo or zebra hair could be found. Thanks to their tracks, we could see that zebras often came here from great distances, just to rub and scratch themselves against the termites' nests. In zebra territories, it is clear that such termitaries play an important part in grooming these handsome animals' striped coats.

On my return from Africa, I immediately had an artificial termites' nest, made of a suitable cement mixture, fitted up in the zebras' cage. When, for the first time, we opened the stable doors, the zebras made straight for this termite castle, rubbing so hard against it that they upset the whole contraption. This was a clear enough indication that I was on the right track. A new termites' nest was at once erected, reinforced this time, and two attendants armed with whips had to keep off the zebras until this piece of equipment, so vital for keeping their coats in condition, had sufficiently hardened. It has been in daily use ever since.

We do not need to go as far as Africa, however, to learn about typical internal arrangements in animal territory, since these can also be found among our native animals ; deer, for example, and rats. A most important feature in the deer's territory is the wallow hole, or mud bath, which is hollowed out by the deer themselves with their forefeet and, after generations of use, may eventually widen out into a sort of pool. Wallowing in mud, like most other activities, is not indulged in simply at the whim of the individual concerned, but is an obligatory specific characteristic, similar for instance to providing stores, or defaecation places, or laying down regular tracks, about which we shall have something to say later.

There are wallowers and non-wallowers. The red deer belong to the former, that is, to those animals in whose territory is included a wallow hole ; roe-deer on the other hand are not wallowing animals, and there is no wallow hole in their territories. We must differentiate clearly between wallowing and bathing, for this frequently takes place at quite a different part of the territory, though sometimes alongside it, as in the Indian rhinoceros.

Another most important locality in the deer's territory is the fighting ground, where, in the rutting season, rival stags fight it out. These special places are not just used for this special purpose for years, but for generations. It is quite untrue to assert that there are no traditions in the animal kingdom. There is very definitely an extremely strong tradition in space pattern, as is shown by the age-old salt licks, those places exceptionally rich in minerals that are constantly visited by animals to satisfy their salt hunger.

Many badgers' earths, and certain animal tracks trodden out over hard rock, are known to be centuries old (Neal, 1948). It is now part of the substance of our well-established knowledge of the lives of animals that most of them are confined within a regular network of localities, where for generations they have wallowed, bathed, fought, grazed, mated, slept, *etc.* Animals live in a fixed space-time system, *i.e.* in a pattern of fixed points, at which they perform definite functions at definite times (Hediger, 1950).

A particularly important type of locality in the animal's territory should not be overlooked, the so-called demarcation places, found with deer, and many other mammals. These exist usually on prominent twigs or branches, or tree stumps, or stones, to which the owner of the territory applies its own property marks, so to speak, in the form of a self-produced scent. We must remember that most mammals are macrosmatic, *i.e.,* they have a literally superhuman sense of smell, by means of which they recognize faint traces of scent, which are quite beyond our powers of detection, as conspicuous signals.

Whilst human beings usually demarcate their buildings and homes optically by means of signboards and street numbers, macrosmatic animals naturally use scents. These are produced in parts of the body varying with the species concerned. In deer and among antelopes, the gland above the eye, the so-called antorbital gland, produces a strong-smelling, oily substance, a small quantity of which is rubbed off on to branches and the like. In this way the whole living space is virtually impregnated with the individual scent of the owner. Any other member of its own species is thus warned off by these scent signals, as soon as it enters an occupied territory.

In recent years many skin glands, whose function it is to provide scent for territory marking, have been discovered. We can clearly see in the zoo how extremely important these marking places are for macrosmatic animals. When animals of the same species are put together for the first time out of neighbouring enclosures, they do not as a rule look at each other to start with, but an examination of scent marks first takes place. While they are inspecting unfamiliar scent trails with their noses, the animals will often brush past each other. Only when the new area has become familiar olfactorily, does optical recognition of the inmates occur.

I will not go so far as to say that impregnation of space by scent—a kind of nest scent—plays a certain part with man as well, although much might perhaps be argued on this score. Monica Holzapfel-Meyer (1943, p. 28) is no doubt on the right lines in her summary of her own valuable research into animal psychology, as follows : " With regard to his division of space, man has remained very close to the animals. We probably imagine that to divide up our homes into bedrooms, dining rooms, drawing rooms, and kitchens is an achievement of civilisation, or culture. Even when there is only a single room, man will make similar divisions and will stick to them with the same tenacity as an animal. The urge to occupy definite fields of activity is so obvious that we only notice it when it ceases. This is sometimes the case in certain mental conditions, when the system of spacial constraints breaks down, giving way to a complete lack of discrimination between places of activity."

I had never properly realized the basic resemblance between man's way and animals' ways of dividing up the earth's surface until on a flight over densely-populated Central Europe. The vivid chess-board pattern of arable and pasture was disconcertingly like the bird territories, as mapped out by Elliot Howard (1920), the pioneer of modern territory research thirty years ago.

And when over Central Africa, the plane flew high above isolated negro villages, the narrow tracks leading to the water holes, or the paths communicating between separate groups of stockaded huts, looked like nothing so much as animal tracks through the empty bush. Man is close to animals, not only as far as so-called fixed points, *i.e.*, locations appointed for definite activities, are concerned, but also in his lines of communication, the tracks which link these together. We may watch a rat or a rhinoceros in the wild as it hugs its familiar tracks, and be struck in either case by the resemblance to a railway carriage running along its fixed rails.

Even in the huge bustling modern city, as G. Hinsche (1944) rightly points out, there are unmistakable counterparts of the animals' tracks. Anybody who today walks to business or to school will agree—on reflection—that he normally follows a definite route, always crossing the road at a certain spot, and preferring one particular side of the pavement, and so on, just as an animal in the zoo or in the wild favours certain little tracks, and gradually tramples them out into paths.

This remarkable reluctance to depart from the normal daily routine is strikingly illustrated in the zoo, during the morning rounds of inspection of the director and his assistants. This daily visitation, along the winding paths, past all the outer cages, through all the animals' houses, and the service rooms, takes up two to three hours. Naturally, it soon becomes automatic, so that one has practically to tear oneself away from the usual route to make a fresh variation. If, by way of exception, one does so, interesting facts in the field of human psychology emerge. Certain keepers may suddenly be encountered, no longer hard at work on cleaning operations, as was invariably the case on the routine tour, but perhaps having a quiet chat, or enjoying the illustrated magazines, or doing other equally astonishing things. As an animal psychologist, such things always remind one of the strength of attachment to habitual tracks !

Hinsche indulged in a scientifically useful bit of fun when he questioned more than 800 schoolchildren, aged ten to seventeen, about the details of their journey to school. As it turned out, most of them not only followed quite a definite path, for instance keeping to the right (or left) of a pillar box, walking under the projecting eaves of a roof, or over a man-hole cover, *etc.*, but many indulged in little mannerisms at particular spots ; for instance, touching a certain post, or stepping carefully over a crack in the asphalt, and so forth. If they failed to observe these rites of the road scrupulously, they thought that it would bring bad luck, *e.g.*, low marks at school.

Only the familiar route made them feel safe and at ease ; the animal, too, only feels safe in its traditional tracks. It has actually been experimentally shown that rats were able to reach safety much faster on their own network of tracks, when suddenly attacked by cats or dogs. On unfamiliar ground they were usually lost. Thus the primitive attachment of the animal to its own tracks is still clearly evident in man's behaviour. In fact it plays a considerable part, as Hinsche has shown, in our understanding of certain diseases of the human mind, From comparative observation, or even from a study of animal psychology in the zoo, we can, as scientific experience shows, gain some insight into a phenomenon so elementary yet socially significant, as sublime yet as primitive as nostalgia in man—that inspirer of the highest poetic creation as well as of criminal action. For it is a phenomenon that cannot be understood at all with reference to man alone. Here the viewpoint of comparative psychology taking into account animal behaviour, is indispensable.

In my book *Wild Animals in Captivity* (1950, pp. 31 *et seq.*), I drew attention to the importance of the quality of space compared with its quantity, usually all too often neglected. The most important part of the animal's territory, the focus, and in the truest sense the starting point, is the home, the nest, the place of maximum security. For human beings too, especially for children, this place is of the utmost importance.

There are two types of adults, as far as attitude to the home is concerned ; the attached and the independent. The latter presumably include those who travel more often and—sometimes in the physiological sense—better than the stay-at-homes. Among individual races and nations there are obvious differences of this kind, for example, the Belgians, whose reputation as home-lovers may have some bearing on their colonial affairs.

Judging by his whole mental make-up, the human baby is just as dependent upon a home, in the biological sense, as an octopus or mouse, a fox or hippopotamus. In their psychotopes, a home is a basically important element, so much so that the lack of it may give rise to serious deficiency symptoms. In the literature of psychology, mental hygiene, and psychotherapeutics, there are exhaustive studies of the uprooted human being (waifs and strays). The reader is here referred to M. Pfister-Ammende's investigations (1950).

The secondary creation of a home is extremely important for those human beings, especially children, who have lost their natural homes, as happened on such a terrible scale in the Second World War, and have thus been the victims of uprooting with all its attendant evils. Those generous charitable organizations which took over the rescue work of children of this sort, noticed repeatedly an extremely significant fact. The well-meaning care given to these poor children in beautiful and spacious community rooms, turned out to be insufficient. All kinds of psychological deficiency symptoms kept on appearing, until the idea of dividing up the room, advocated among others by Professor S. Bayr-Klimpfinger, was adopted.

It consisted, in short, of splitting up the spacious day-rooms into separate territories, and in particular of making homes inside these compartments, within which the children could in a real sense feel at home. The Pestalozzi villages—conceived and first carried out by Walter Robert Corti in which, under optimum conditions, refugee children from all over the world can feel happy and settled in a home, in a familiar community, represent, both humanly and biologically, an excellent and ideal form of the successful artificial creation of a home. K. Heymann (1943) has vividly described how human children—like animals—often identify themselves with their house, *i.e.*, with their home. There are of course animals that, in varying degrees, are actually physically connected with their houses, or whose bodies or bodily parts form offshoots of their homes. The rim of the limpet's conical house fits tightly over the spot chosen as its home on the rocks. The hermit crab's powerful claws act as a security precaution for the house which its body fits so perfectly. Many other animals have organs—often a specifically shaped head or tail—that act as a door. This phenomenon is known as phragmosis. The wart-hog (*Phacochoerus*) is the biggest example of this kind of thing. At night, it lives in underground tunnels of twelve to eighteen inches diameter. When alarmed, it dashes back to its home, swings quickly round at the entrance, and backs in, so that its powerful tusks project towards the entrance, forming a secure defence.

Too little attention has been paid to an important factor in the compulsory sharing of living quarters—both in the internment of human beings and in the captivity of animals—that is, to their psychotope, formerly called archetope. It has long been assumed that an animal is tied to, and even within—its living space, as far as its structure and activities are concerned ; *i.e.*, morphologically and physiologically (ecologically). In the mental field, a similar kind of adaptation to this biotope may be observed ; a tuning in to the psychotope. When Russians from the spacious flat steppes were interned in a sub-alpine region in Switzerland, which we regard as a delightful holiday district, their reaction was marked, as M. Pfister described (1949). It transpired that these men from the boundless West Asian plains felt unhappy among our mountains. To them, these seemed sinister, oppressive. A similar eerie feeling of being oppressed, eventually becoming intolerable, is familiar to Europeans forced, as geologists, or colonial officials, to penetrate into the trackless primeval forests of Central Africa. After a few days, even thoroughly well-balanced men usually feel more or less depressed, and only regain their spirits when they see the light of day, or reach open paths. The opposite is true of the pygmies. For them, the open landscape is sinister and unbearable. Their psychotope is the thick virgin forest. Even those accustomed to continuous contact with white men move through the forest when going from place to place, if they can. Many animals clearly show strong psychic attachment to particular country, the psychotope. Here is a further instance of similarity between the spatial experience of men and of animals.

Another resemblance emerges, when one studies the distinguishing marks and boundaries of the living space, *i.e.*, its territorial demarcation. While the individual, when defending its territory against the intrusion of outsiders, is directly concerned with acoustical, and especially with optical, marking, new possibilities arise with olfactory demarcation of territory. Scents, whether of dung, urine, or glandular secretion, are detachable from the body. They can literally be separated to act as place-reservations. The particle of dung, or the trace of secretion on the marking place, becomes, as Bilz shows, the *pars pro toto* (part for the whole), and continues to be efficacious even in the absence of its author.

Bilz, discussing the demarcation behaviour of bears and dogs (1940, p. 285), states: " Excrement and the image conveyed by it, frighten and even terrify the intruder. Is not this similar to magic ? That tree-trunk scarred through being rubbed is still potent, even though the bear fell a victim to the hunter's gun weeks ago, and the scent banner unfurled against a tree stump by the dog continues to strike terror, even when the animal has long since changed masters and gone to live on another farm. Bear and dog alike have laid a ban on the district, to show that it is their home. Cave ! Taboo ! "

It is likely that scent marks lose their effect after some weeks—that is why they are sometimes renewed by the animal once a day, or even oftener. Bilz's comparison of the scent-marking of territories with primitive man's magic and taboos seems to me significant. Among South Sea Islanders, and many other races living in a state of nature, practices occur that basically represent ownership tokens, such as marking freshly-cleared ground in the primeval forest, or the recognition marks applied to his kill when the hunter cannot take it with him at once. In many cases, it is not only a question of objective marking, but of laying a spell, or putting on a taboo—of defence measures against rivals and demons. Even as near to Europe as Morocco, I came across Berbers buying a tarry "secretion" from the medicine man in the bazaar, which they took home in short lengths of sheep's intestine, and applied to all four corners of a hut or house to ward off djinns, those abundant evil spirits. As with flight behaviour (cf. Chapter 4), so with demarcation behaviour in the human subjective world, the dangerous larger animals seem, in the course of development, to have been replaced by dangerous demons, and defence behaviour against animals gradually seems to have become superfluous, and have been correspondingly transferred to demons. Out of defences against rivals and enemies, magic was born. I cannot leave this chapter on animals' homes without drawing the attention of the reader who may wish to go more fully into the subject to David Lack's excellent and at the same time charming book, *The Life of the Robin* (1946). It not only contains a complete account of the robin's territory, as well as of the life habits of this bird, but is at the same time to be recommended as a most pleasing introduction to animal psychology.

3

THE ANIMAL'S DAILY LIFE

CONTINUING our plan to give priority to simple and appropriate questions of animal psychology we now turn to the equally simple question—what does the animal in freedom actually do all day, or all night long ? This simple question, surprisingly enough, has only recently been asked by animal psychologists. At first, interest was focused mainly on the distribution of phases of rest and activity within a 24-hour cycle, since of course it was clear that many animals are active only by day, others only by night, while others again partly by day and partly by night as well.

In particular J. S. Szymanski (1920), the psychiatrist and animal psychologist, working for a time at Basle, and who later (1921) disappeared while on a journey to Vienna, devoted himself to examining the distribution of activity and rest in various animals. He took so-called actograms of everything, from snails to canaries, from goldfish to human babies. For these he used simple graphs consisting of two concentric circles. The circumference was divided up like a clock face. At the top, the figure 12 marked midday; midnight was at the bottom; 6 a.m. on the left, and 6 p.m. on the right. On the inner circle, the rest periods were shown by thick lines; on the outer circle, the periods of activity. By means of this diagram, the distribution of rest and activity of any animal can be clearly shown.

Naturally, the correct values had at first to be obtained; *i.e.* each animal had to be watched to find out when it was quiet and when it was active. Szymanski invented a whole series of apparatus for this purpose—small cages so delicately hung that the slightest movement of the inmate was transmitted to them. A sensitive needle then recorded on a kymograph, or paper-covered recording drum, each change in the centre of gravity. The baby with its cot was put into this kind of cage; the goldfish had a fine thread tied to its dorsal fin, leading over a small pulley with a counter-weight, and then fastened to a delicate recording needle.

This apparatus did not prove ideal in all cases, yet with its help a valuable beginning was made. It showed for example that there are monophase and polyphase species. Those with only a single period of activity and a single period of rest in 24 hours, such as the canary which sleeps all night and is a typical diurnal animal, are monophase, as are most monkeys. The crab, rat and rabbit however are polyphase. In the daytime, they occasionally indulge in intervals of rest, and are also active at night. They change over from states of rest to states of activity, and *vice versa*, several times within 24 hours.

Szymanski's experiments are invaluable, it is true; but they tell us nothing about the different kinds of activity of the animal. His results are purely quantitative, and give us no details about the animal's occupations all day long. Yet it is extremely important for us in the zoo to know how the ordinary day of an animal appears under natural conditions ; which are its chief and which its secondary activities; for in the zoo we would like to cater for the animals, in this respect, as naturally and as suitably as we can.

The *qualitative* analysis of the animal's daily life was only studied a few years ago, which is not hard to understand, since it is much more difficult than the quantitative analysis begun 30 years ago. To do this, it is not enough to keep an animal in a recording cage. Normal occupations can only be observed in the natural surroundings of undisturbed life in freedom. And so much closer, more detailed work was here required from the field biologists—just as was the case in research on animals' homes. Naturally their results could usefully be checked, completed, and developed in the zoo.

Anyone who carefully watches an animal, a mammal let us say, in wild life will immediately realize that it is no easy matter even to get near to the animal, since it is always on the look-out to prevent anybody (*i.e.* an enemy) from coming too close. In fact, this constant watch for enemies that may threaten it at any time and from any direction, is the wild animal's chief preoccupation. It is ever on the alert, so as to avoid enemies and be ready for escape. This perpetual, never-ending activity, even during sleep, represents so fundamental an occupation—overriding all other behaviour— that we must devote a special chapter to discussing it.

Before turning to the various possible kinds of animal activity, we will first examine the resting phase, especially sleep. It is not quite correct, as a matter of fact, to regard sleep as the purely passive state, in contrast to the other activities, as, in the biological sense, sleep too is an active function of the organism. According to N. Tinbergen (1951), sleep is a true instinctive action, preceded by " appetenz " behaviour, such as seeking a special sleeping place, assuming a particular posture, *etc.* In addition the sleep centre is situated in the hypothalamus, as are the centres of other instincts.

True, there is an extremely useful amount of laboratory research, for instance, on sleep centres of the brain, or on the character of hibernation. Yet we still know very little about animals' sleep habits; *e.g.* normal duration of sleep. The experimental cages that record changes from rest to activity just mentioned are no good for our purpose, for the simple reason that this apparatus makes no distinction between complete rest and true sleep. It is impossible to tell whether many of the lower animals, such as fish and reptiles, are asleep or awake.

Most animals have special sleep postures and, as we have said, special sleeping quarters, but it is very rare to find them in this typical situation, because all wild animals at once react to the approach of their enemy, man, by taking to flight.

Certain fish, such as carp, electric catfish (*Malapterurus electricus*), and some cichlids (*Astronotus ocellatus*) *etc.* lie at the bottom of their tank at night, so that we are justified in saying that they sleep. Bats sleep by day, hanging head downwards, as is well known. Some species of seals—for example sea lions and common seal—can sleep under water, only coming to the surface periodically to breathe. Many mammals, as for instance the fox, use their bushy tails as pillows, and others curl up into a ball. Flamingoes and other birds are in the habit of perching on one leg while asleep. This is not the place, however, to go into all the various sleeping habits of animals.

I should like to deal only with elephants' sleep habits in greater detail, since until a few years ago they were subject to some very queer notions, and I was able to do some research on them, both with Indian and African elephants. Both have enviously little need of sleep. Adult animals, as a rule, do not go to sleep before midnight, and then usually for only about three hours. Elephants that are very old, unwell, or upset, do not lie down to sleep at all, but stand up all night; while healthy animals lie down at full length, with their trunks coiled up like a ship's rope, as Benedict described it (1936). Elephants' sleep was for long a mystery, because they sleep so very lightly. In practically no zoos can elephants be watched sleeping, as they wake up and trumpet at the slightest disturbance, or unfamiliar noise. In spite of innumerable carefully prepared attempts, I was never fortunate enough to catch any of our old elephants asleep at Basle Zoo. Even when I left their door unlocked, to stop any sound of lock and key, and crept in on tip toe, I invariably arrived after the animals, clanking their chains, were already standing up.

In the circus, however, where the elephants were used to all sorts of noises at night, including the night watchman's continuous visits, I was able to see the elephants asleep whenever I felt like it. The fact that in the summer of 1944 Knie's Circus had fifteen Indian elephants in its menagerie presented me with a wonderful opportunity of checking over and adding to the few facts known about the sleep habits of these huge beasts. During the night of my experiment the tent was illuminated until 11.45 p.m., when the light was switched off. Only a small paraffin lamp now cast a faint gleam on the fifteen powerful creatures, chained to their wooden floor in a straight row. In the dim light, the fifteen different physiognomies looked like enormous masks.

A few minutes before midnight, all the elephants were still standing up, twisting the hay into wisps with their flexible trunks and feeding contentedly. Some threw the hay and straw on to their heads and backs with typical movements, so that they looked a comic sight. On the stroke of midnight, two animals that were standing side by side, " Sandry " and " Rosa ", lay down together on their left sides with surprising ease. Four minutes later, " Tabu " followed suit. At twenty-seven minutes past, old " Mary " settled down and so did her neighbour, " Countess ", one minute later. Both lay on their right. Soon the notebook was filled with minute details about every single elephant. One or two hours later, most of them had to stand up to relieve themselves of the waste products of their enormous digestive systems. That night, with the manager Rolf Knie as my tireless assistant, we made the most detailed record of Indian elephants' sleep that has ever been taken. Their average length of sleep was two hours nineteen minutes. This figure agrees remarkably closely with the two hours fifteen minutes that the American physiologist Francis Benedict recorded in 1936 for a single elephant, averaged over a space of several nights.

Two of Knie's elephants, " Claudi " and " Punschi " did not lie down at all during the night when they were under observation. Perhaps in Punschi's case it was due to the fact that he had been under the dentist that day. No reason for Claudi's behaviour was apparent, though it later transpired that she must have been seriously ill at the time, as she died of tuberculosis soon afterwards. Probably both these animals slept intermittently standing up, as elephants are known to do, though this sort of sleep apparently never lasts for more than fifteen minutes. Standing sleep, which also happens with elephants in wild life, can be recognized by quietness, the cessation of ear flapping, the limply hanging trunk, and the closed eyes.

When elephants sleep standing up they show a marked tendency to support their trunks on something. We often saw animals in the circus resting their trunks on their comrades asleep on the floor without in the least disturbing them by doing so. It sometimes even happened that a waking animal suddenly rubbed his hindquarters hard against one of the elephants lying asleep. Punschi was sometimes jammed by her sleeping neighbours so tight that she did not know where to put her feet, so she set one foot firmly on her sleeping, right-hand neighbour's head, close to its eye.

One of the most surprising results of this night spent with Knie's elephants was that these great beasts would suffer the most severe disturbance from their fellows without ever being wakened up, while they immediately started up from sleep at the least disturbance from human beings. This fact proves that the central nervous system functions properly during sleep. But the incoming stimuli are filtered, so to speak, *i.e.* divided into biologically harmless ones from the members of their own species, and potentially dangerous ones from human beings or other enemies.

Four years after this, I had the opportunity of comparing the sleep habits of the Indian elephants with those of the African, in the only state-owned farm for taming African elephants in existence, at Aru in the Belgian Congo. This equatorial night spent watching elephants was far longer and more tiring than the one in the tent in Switzerland at midsummer. Some fifty African elephants of all sizes were tied up in the open air, like ships in a harbour, in two parallel lines. A number of them were kept under constant observation until dawn. It grew dark shortly after 6 p.m. Every step I took, I was accompanied by a native bodyguard armed with a spear, for no one was ever certain whether an elephant might not break loose, or a strange wild elephant find its way into the animals' ranks. A second group of natives lay round a small fire on picket duty.

By 9.29 p.m. a very young animal had lain down, and it then slept for more than four hours altogether. Four others sank down asleep on the soft ground at 10.30, but most of them did not start to drop off until after midnight. Neighbours often lay down together. In the case of elephants with full-grown tusks, this was not easy ; they had to twist their heads right to one side in order to fit their gleaming ivories in. Before lying down, some of them first took a nap, supporting their great craniums on their tusks, which reached down to the ground.

As with the Indian elephants, preparations for sleep are made while they are standing. The moment the elephant touches the ground as it lies down, it falls asleep. One can tell this by the typical variation in breathing which becomes deeper in sleep, strikingly regular, and much noisier. If one inadvertently trod heavily, or rattled the pocket torch, the colossal beasts at once shot up snorting, and would not lie down again for hours, and sometimes not even then, so that all our trouble was in vain.

Some of the animals made a sort of pillow out of the piles of papyrus stems, lemon-scented grass, and twigs that their native keepers had heaped up in front of each elephant for fodder shortly before dusk, just as their Indian cousins in the circus had done with the straw. In other respects too there was a remarkable similarity between the sleep habits and length of sleep of the Indian and African elephants.

While watching the elephants sleeping beneath the cloudy sky, the question as to whether they dream seemed even more insistent than in the case of other sleeping quadrupeds. It is true that I could find no real evidence of it, but in view of other observations, we must not assume that dreams are an exclusively human prerogative. Serious difficulties arise in studying dreams, even in human beings ; the study of animals' dreams must therefore present even greater problems, in the absence of two extremely important aids, namely, self-observation, and interrogation. Moreover, we must take into account the fact that an animal's sleep is often of quite a different character from man's, and that we are still very ignorant about animals' sleep. So the prospects for the study of animal dreams are, to say the least of it, very unfavourable. Even the question as to whether animals do in fact dream is difficult to answer.

It is justifiable to assume that dreams occur in higher animals, since the brain, the nervous system, and especially the whole body of the sleeping creature, continue to function. For example, during hibernation, when the frequency of heart beats and breathing, body temperature, and therefore reaction capacity, are at a minimum, dreams are as little likely to happen as in the periods of hibernation of reptiles, and other cold-blooded animals. Dreams are inconceivable, except in the normal restful sleep of animals, so much like our own sleeping state.

Before turning to study an animal's dreams, we must first of all see if we can find sleep more or less resembling man's somewhere in the animal kingdom ; for then only can we hope to find dreams. There do in fact exist similar conditions of sleep among warm-blooded animals, that is, mammals and birds. Whether they also exist in the case of the next lower order, the reptiles, is doubtful. We can for instance assume, with a fair degree of probability, that a lizard lying motionless with its eyes shut, is asleep, provided that it is in a milieu of optimum, or sufficiently high temperature, so that the state of hibernation, so easily induced in cold-blooded animals, can be safely excluded. But not all lizard species can close their eyes; this faculty is absent for instance in members of the species *Ablepharus* and *Ophiops*. In their case, both eyelids are fused into a transparent cover, shaped like a watch glass. The same thing happens in the case of snakes' eyes. That is why we get the impression that they always keep them wide open. For this reason, it is extremely difficult, if not impossible, to tell whether a snake at rest is awake or asleep. Some years ago, in 1937, the American reptile expert R. L. Ditmars believed he had discovered that the sleep state of a snake could be told by the fact that the pupils did not lie in the centre of the eye, but were turned down to the lower rim. Yet in spite of years of study, and of sifting a large amount of material on snakes, I have never managed to confirm this observation, nor to find it confirmed in the literature.

It is just as difficult to state anything definite about sleep in fish, which are also unable to shut their eyes and which, during sleep, if that is what we can call it in their case, keep up a continuous slight motion of the fins to maintain position. Not all fish are in the habit of lying still on the bottom at night, as carp do, or of floating gently along at the surface, like the giant sunfish (*Mola*), which can be caught by surprise, if approached quietly enough. Many animals, that are known for certain to sleep, make continuous movements during their sleep to keep themselves in the right position ; this is so for certain water birds which drift along the surface during sleep, as well as for certain mammals that sleep on the surface of the water, or that have to surface from time to time to breathe, such as some seals, sea otters, *etc.* The Basle zoologist, Friedrich Zschokke, says (1916) that a sleeping duck makes continual paddling movements in the water with one leg. " It keeps on describing circles like a boat being rowed by one oar, and thus escapes the constant threat of shipwreck on the shore from winds or currents." Such behaviour habits show that not all movements performed during sleep should be interpreted as dream symptoms.

Even preliminary questions of this sort about ability to sleep in animals present difficulties, as we have seen, when it comes to stating which animals can sleep, in the human sense. The question about whether animals dream is even more ticklish. There are really two questions here. First, " Is the raw material for dreams present ? " Second, " Do the indications present point to the fact that this raw material is used ? " If either of these related questions is to be answered separately, it follows that, as far as the higher animals are concerned, they can be answered in the affirmative.

The raw material for dreams consists essentially of memory and imagination, and animals clearly have more or less good memories, as has been proved dozens of times by careful experiment, this corresponds exactly to our everyday experience. Some animals in fact possess quite outstanding memories, and there are experts who think that the fox's or the horse's memory for places is far superior to man's. Thus the animal does have the faculty for remembering past experience, a fight perhaps, or a meal, based on certain stimuli or memories. At the same time, there is no doubt that a certain amount of imagination exists as well. The Dutch animal psychologist, F. J. J. Buytendijk (1933) mentions the so-called " vital phantasy " of the animal. To the cat playing, the paper or rubber ball is not a dead object but something alive, perhaps even its prey. Thanks to its " vital phantasy ", the animal sees all sorts of hidden properties and potentialities in its plaything, inviting it to movements of investigation, and perhaps leading to play. We must therefore allow that the raw material for dreams is present in the higher animals.

Let us now turn to the second part of the twofold question; whether this material can be used, *i.e.* whether sleeping animals sometimes show behaviour habits that indicate dreaming. A further basic difficulty now appears ; that is, how to approach the sleeping animal unnoticed, and without disturbing it.

The wild animal, with its marked tendency to escape, is notorious for the fact that it is never completely released from that all-important activity, avoiding enemies, even during sleep, but is constantly on the alert. Its sense organs are not inactive, but its central nervous system sorts out the stimuli it receives into dangerous ones, *i.e.* coming from possible enemies, and harmless ones. Only dangerous ones set off the alarm, and then the animal starts up, and runs away if necessary. Naturally, when the human observer—a possible enemy—approaches, the wild animal's alarm goes off, the sleeping animal wakes up, and vanishes. Even thoroughly tame wild animals, for which man has long ceased to be a dangerous enemy, seldom remain in their characteristic sleeping position at the approach of their trusted keeper. Almost without exception, they wake up, giving one hardly any chance to watch them fast asleep for any length of time. One is always impressed by the practical impossibility of getting near enough to take a photograph of a sleeping wild animal, in a typical sleeping position.

And yet there is one group of animals that have largely, even completely, lost their tendency to escape from man, and in consequence do not allow themselves to be noticeably worried by man's presence. These are the domestic animals, especially dogs and cats, that live in close contact with man night and day. Naturally they are most easily watched when asleep, and thus we have perhaps in their case the best chance of studying dream symptoms. So it is not surprising that the oldest, most numerous, and most complete data on animals' dreams are gained from the dog, man's most intimate domestic companion, the oldest of house animals. We find such observations even in Lucretius, the pre-Christian Roman naturalist, in the fourth book of his famous *De Rerum Natura*. Ever since, the dog has been the favourite, in fact almost the only, animal subject of dream investigations.

Carl Gustav Carus (1779–1868) states in his *Comparative Psychology* that all mammals and birds often dream, the dog most often of all. Carus made a study of his Newfoundland dog " Sultan ", and thanks to his accurate knowledge of the animal, its way of moving, and the changes in its voice, believed that it was quite " easy to follow what sort of dream-pictures were occupying its mind at any particular moment." Here Carus clearly failed to take into account the great risk inherent in this method—that a far too complicated dream content can be read into trivial twitching, or noises, without any possibility of testing it. Carus considered as particularly significant the muffled, distant, barking sounds of sleeping dogs. " Anyone who has listened to people talking during sleep, or in the first stages of somnambulism, must have noticed a distinct resemblance between such strangely altered speech, and this dog's oddly changed voice, coming through jaws almost completely closed."

The Italian doctor, Sante de Sanctis (born in 1862), made a more critical analysis of animals' dreams. This author also assumes that dogs dream, and regrets the lack of accurate observation on this subject. Among the many authorities he quotes in this connection, he mentions Fr. L. Goltz (1834–1902), who examined the problem experimentally. Thus Goltz completely removed both cerebral hemispheres from a dog ; the animal victim lived for a year and a month, sleeping all the time, " but making no movements during sleep which might have led one to infer that it was dreaming."

In order to solve the problem about dreams, Sante de Sanctis used the method of interrogation. He gathered information from numerous hunters and animal breeders, to find out whether they had noticed any sort of behaviour that might denote dreams. The results of the many enquiries led him to the conclusion that not only do dogs of all breeds dream, but all the higher animals, mammals and birds. This much is certain, says de Sanctis at the end of the chapter on animal dreams, " that man's dreams and animals' dreams do not differ essentially, but only in degree." This investigator was certainly the first to propound the question as to whether an animal knows that it is dreaming. He came to the conclusion that the animal was incapable of distinguishing between the dream and the reality, so he thought that the animal mistook the dream experience for the reality, as sometimes happens with man under certain conditions, especially pathological ones.

It would certainly never occur to anybody to deny that a baby has the power of dreaming, just because it cannot tell us anything about it. In his pioneer work on *The Mental Development of the Child* (1922), the psychologist Karl Bühler states that the earliest phases of the child's conceptual activity, and thus the problem of dreams in babies, must be studied by purely animal psychology methods. According to Bühler, sleeping babies reveal a state similar at least to dreaming in their very first months, by gestures such as distinctly smiling, audible whimpering without cause, and sudden cries that usually wake them up. A. Peiper (1949) thinks it is impossible to tell when dreams first occur in childhood. He quotes Hoche, on the other hand, according to whom children start dreaming very early, certainly at the age of two.

Very similar behaviour, says Bühler, sometimes occurs in animals. Anyone who has watched a sleeping hound twitching, often becomes convinced that the animal is dreaming of the chase ; its whole muscular system is somewhat tense, its head and paws work in characteristic manner, and it emits whining noises. If its forepaws are gently touched, it sometimes snaps its jaws, as if seizing its prey. The same thing may happen without any external stimulus; or the dog suddenly starts up awake, just as we wake up with a start from a vivid dream. We have no reason—says Bühler—to consider such observations on animals as essentially different from those on sleeping human beings. The dog, too, dreams, *i.e.* events take place within it of perception-like or hallucinatory nature and of great vividness, repetitions of things it has experienced, more or less recently, when awake. In his book, *Pars pro toto* (Leipzig, 1940), the neurologist Rudolf Bilz considers that during sleep the dog betrays a distinct inner life. " It seems as though the sleeping animal rehearses its daily life in pantomime. We can hear it growling or smacking its chops, we can even watch it making running motions, though often only sketchily—*pars pro toto*—expressing hunting, or escaping, in a rudimentary show. We can hear it apparently suffering from nightmare, and groaning. Is it being threatened by invisible enemies in an inner world ? We can also watch it expressing pleasure by wagging its tail while asleep."

31

Most other investigators who have hitherto dealt with the problem of animals' dreams hold similar views. Thus Friedrich Dahl in his *Comparative Psychology* (1922) considers it very likely that the higher animals at any rate dream in a similar way to man. E. Kretschner (1930) sees in man's dreams a relic of former hereditary stages of development of our mental life. Bierens de Haan (1929) considers it very probable that the higher animals at least dream in a similar way to man.

We are indebted to Dr. Rudolf Menzel (1937), the expert on dogs, for fundamental research on early signs of dreams in growing hounds, which has gone on for over sixteen years and during which he has studied about a thousand dogs, especially boxers. Menzel distinguishes various dream types: feeding, snarling, barking, and running. In a six-day old puppy he was able to observe its first food dream; it consisted of a variety of lip smacking and sucking noises, together with motions of its body while it slept. Snarling and barking during dreams were first recorded on its tenth day. Scolding, with continued running and feeding motions, was observed in a 24-day-old puppy's dreams, after it had had an exciting day, when it visited the scientist's study for the first time, and where it had at last fallen asleep by the fire, exhausted after an extended playtime.

While the character of Menzel's observations on dogs' dreams is purely descriptive and documentary, Erhard, and above all, Sarris, have also attempted to solve experimentally the problems about dreaming in animals, especially dogs. H. Erhard (1942) noticed that his dog hunted in dreams ; it barked in a high pitch, twitching its legs, and kicking out. This always occurred, as a matter of fact, after the dog had just been taken for a walk in the woods. If the dog was then wakened, it showed no desire to go out, which would certainly have been the case, thought Erhard, had it remembered the dream. If the dog had not been in the woods for some days, " hunting dreams " could at once be induced by giving it artificial pine-needle scent to smell. The author thought the smell of the woods was associated with a memory image of hunting. One ventures to wonder whether the artificial pine-needle scent had the true characteristic scent of woodland surroundings, to the superhumanly sensitive dog's nose, and whether it would not have been more adequate to have tested how the dog reacted to game scents separately presented.

The dog expert E. Sarris (1935) used biological stimuli in the form of bones and sausages to make his bitch " Leon ", on which he was experimenting, dream. Sarris, too, is convinced that the animal can re-live important events of daily life in dreams. After a meal of bones, which the animal had not enjoyed for some time previously, the sleeping dog's jaws moved in such a way as to suggest that it was trying to chew something hard. A few seconds later the dog gulped. Chewing and swallowing lasted exactly two minutes ; the animal then lay quite still for three minutes, and eventually, sighing, opened its eyes.

In another experiment, Sarris laid a piece of sausage at 12.30 p.m. an inch away from the nose of the sleeping dog. In its waking state, the dog would easily have noticed the sausage at about two yards. After thirty seconds, the dog began to chew, then opened and closed its eyes, and the skin of its upper jaws twitched spasmodically. Chewing, blinking and

Figure 3a.—African elephants lying asleep in the Station de Domestication des Elephants at Aru (Belgian Congo). These snapshots, taken in 1948, probably represent the first pictorial documents of the sleep of these animals, so long a mystery. When it can, the elephant makes itself a large ' pillow '. It is by chance that all three are lying on their right, but the trunk-curling is normal.

Figure 3b.—Animals with big tusks often have difficulty in lying down.

Figure 3c.—Young elephants sleep longer than old ones.

Figure 4.—Big horn sheep (Ovis canadensis) in San Diego Zoo, California, characteristically fighting each other through their dividing fence. During this, the rams stand up
about ten yards apart, then dash towards each other on their hind legs, like fantastic masked
dancers. Even the iron posts of the fence get broken. While the fight is on, the females in
the background sniff at the marked tree, i.e., at their ram's demarcation spot.

Figure 5.—Male wild boar (Sus scrofa) with six or seven month-and-a-half old piglets
climbing or lying on him. It is wrongly asserted that male wild pigs eat their young if they
get the chance, but long zoo experience contradicts this.

twitching lasted three minutes, three seconds. The dog then lay still for two minutes, sixteen seconds. At 12.35 the dog woke up. On first opening its eyes, it did not notice the sausage, which was then removed. Sarris even claims to have observed dreams of a sexual nature in the animal studied.

The method of provoking corresponding reactions by presenting certain stimuli to the sleeping dog should give clear results, it if were possible to make the trained dog carry out, at least to a certain degree, carefully graduated commands while asleep. Up till now, however, as Werner Fischel states (1938), only inborn co-ordinated movements could be recognized in sleeping dogs, never acquired ones. It remains to be seen whether systematic observation of a number of dogs, continued over a considerable number of nights, might not give a similar result.

Fischel thinks, moreover, that he can interpret everything he knows about sleep movements, either from personal observation or written accounts, as so-called " vacuum activities ", in K. Lorenz's sense (1937, p. 298) ; namely, instinctive motions, successfully performed in the absence of external stimuli. But in the case of Sarris's artificially provoked " sleep motions ", for instance—apparently unknown to Fischel—there is no question of a lack of stimuli, such as is typical for the vacuum activity.

Apart from this classic animal of research, the dog, the dreams of other domestic animals have been described, particularly those of the horse Hempelman (1926) quotes a report of K. C. Schneider on a horse which had been in service during the Turko-Italian war, and in later life used to neigh excitedly in its sleep, kicking out with its hoofs, as if it were living battle scenes over again. By day it was calm and docile, but by night was liable to be excited and bite. Th. Kottnerus-Meyer, the former director of Rome Zoological Gardens, mentions (1924) a Hungarian artillery horse, Memmo, that had taken part in the First World War, had various scars from shell splinters, and was nervous and easily frightened. This timidity soon disappeared whenever the horse's blinkers were removed. " The animal could now see everything, and quietened down. Its vivid dreaming was remarkable. During sleep it often neighed aloud against its stable walls. Doubtless this was the result of its war-time experiences, for even a horse cannot emerge unscathed." According to Stefan von Maday (1912), Schopenhauer also believed that horses dream. In his *Psychology of the Horse*, Maday quotes a report from H. Bouley's *Leçons de pathologie comparée* on erotic dreams in a horse under anaesthetics. It was a stallion that had to have a major operation on its hind leg. During the operation the anaesthetized animal suddenly emitted a gentle whinny, which sounded just like the neighing of a stallion let out to the mare. At the same time, it went through movements suggestive of copulation, as far as its straps allowed it.

Remarkably few observations are available on cats' dreams, despite the fact that, after the dog, it is the domestic animal living in the most intimate contact with man, and can often be seen asleep in the day-time. In his *Tierpsychologie vom Standpunkte des Biologen*, Hempelmann simply states that cats sometimes spit during sleep. According to Washburn (1930), Darwin includes cats among animals that certainly dream, along with dogs, horses, *etc.*

Accounts of wild animals' dreams are even less frequent. In his *Natural History of Dreams*, M. H. Baege (1928) does not mention them. At the anthropoid ape farm on Tenerife, Rothmann and Teuber (1915) observed some sleeping chimpanzees that occasionally made movements and noises as though they were dreaming. Authorities, such as W. N. Kellog and L. A. Kellog (1933), who brought up a human and a chimpanzee baby together under identical conditions, studying their sleep habits minutely, and comparing them, have unfortunately nothing to say about the dream behaviour of their charges. P. Scheitlin (1840), pioneer Swiss animal psychologist from St. Gallen, mentions elephants, of which he made a special study, that had vivid dreams. True, he gives no further details of any kind. Bastian Schmid (1930) relates how his tame foxes used to purr and grumble quietly in their dreams. K. M. Schneider (1933) happened to notice young polar bears, that often used to start in their sleep, without apparent external cause, and when licked by their mother while asleep, began to make a humming noise, just as they did when being suckled. This authority would have considered it premature to call that sort of behaviour dreaming.

Dream research in animals has brought forth no surprises nor even appreciable advances in the last few years, or decades for that matter, as this short summary shows. Friedrich Zschokke's interpretation, put forward in 1916 in his work on animals' sleep, still holds good, namely, that in a certain sense animals and birds dream. " Yet the chief problem remains unsolved, that is, whether animals' dreams are comparable with the varied sequence of images which man's mind automatically and uncontrolledly produces during sleep, and which out of nothing invents its ephemeral creations with inexhaustible lavishness. The animal stands perhaps on a lower plane of these wonderful manifestations of the mind, on the threshold of the fairyland of dreams."

Apart from sleep, animals have a special kind of rest, a sort of semi-resting, called dozing. This state of rest is also found among primitive races, but in civilized man, as a rule, there are only hints of it in that intermediate stage between waking and sleeping. Dozing is distinguished by a more delicate filtering of the external stimuli. The dozing animal is far more alert than the sleeping one, and is thus quicker to take alarm. Many animals have a dozing position, as distinctive as their sleeping position —thus, for instance, dozing ducks and geese only tuck their heads under their wings far enough for their eyes to remain uncovered ; in true sleep their eyes are completely hidden.

Typical dozing is also found in ungulates that chew the cud. This type of food digesting often takes place with the eyes half closed, automatically, as it were, and often in different places in the territory from the sleeping quarters. This reference to chewing the cud brings us to another important occupation of the animal, namely the acquisition and assimilation of food.

All the activities still to be examined are much less important than the dominant avoidance of enemies, and may be divided into the six following categories :

1. Acquisition, including assimilation of food.
2. Upkeep of home (maintenance of living quarters).

34

3. Social activity (encounters with its fellows).
4. Reproduction, and care of young.
5. Care of the body, and comfort behaviour.
6. Play.

We often find in the animal kingdom that parts of the body are specially developed to perform these functions. For example, a very high degree of specialization can be seen in organs devoted to the acquisition and assimilation of food, such as teeth, beaks, claws, *etc.* For the upkeep or the demarcation of living quarters, such special adaptations are found, as for example the woodpecker's beak, by means of which a nesting hole can be hammered out in a tree trunk, or the badger's claws, ideally suited for digging, or the many skin glands that provide scent for marking territory olfactorily. Signalling apparatuses and special weapons are used in social encounters, and are exclusively employed for this purpose, and never for fighting animals of other species (*e.g.* the mandibles of bees ; the horns of the giraffe). The presence of organs of reproduction, and the care of the young, are familiar enough to require no further mention. Finally, all sorts of organs for the care of the body have been noted, *e.g.* cleansing claws in the kangaroo, the hyrax, and the beaver, ideally adapted for cleaning their coats.

Only one type of activity, the last in the list, namely play, has had no special provision made for it, significantly enough, in the make-up of the animal's body. In the grim business of existence in natural surroundings, play is not an obligatory activity for the wild animal, but only an optional one, often only fully developed for a short while in youth. Until now, no play organ has been discovered in any single animal, in contrast to the organs belonging to all the activities just enumerated.

Acquisition of food in wild animals does not simply consist of the assimilation into the body of quantities of vegetable or animal food. The particular manner of intake of food is intimately linked with psychological characteristics. Animals of prey and herbivores represent two fundamentally different types, connected however by many intermediate forms. Among insectivores and carnivores again, there are the active and passive types. According to the requirements of one or the other, food acquisition claims a greater or lesser share of the daily life of an animal.

The active type must find its victim by actively searching and tracking, *e.g.* the common toad, the ring snake, the golden eagle, or the ermine. They all have to show considerable activity to find their prey and satisfy their hunger. The passive type is quite different—lying in wait, motionless, until some unsuspecting prey comes within reach. Among these are the pike, the horned frog, the chameleon, and the little heron, and among mammals, the lynx, and certain sneak cats.

Such animals are distinguished as a rule by three groups of features. They are largely invisible, *i.e.* suitably camouflaged ; they have the faculty of acting with lightning speed, once their time for activity has arrived; and on the other hand, their mental make-up permits them to wait motionless for hours on end. All these characteristics are clearly developed in the chameleon for instance. It is often difficult to see it among the bushes, even before one's very nose. It can shoot out its sticky tongue truly explosively at an insect, a locust, or a praying mantis, and can be seen squatting motionless in the same bush for days or weeks. This sort of waiting about for food takes up a great deal of its daily life, while a large carnivore or bird of prey can sometimes kill enough food for a whole day in a matter of seconds.

Let us consider another of the six chief occupations of animals on our list, namely, the care of the body, and the so-called comfort behaviour. These activities take up a considerable time among the higher animals, especially mammals and birds. Such birds as ducks and pelicans may spend hours preening their feathers. One by one, each feather is drawn through the beak, smoothed, and set in order. Vultures, marabou storks, cormorants, *etc.* stand continuously in the sun with outstretched wings. Mammals often spend a long time on their coats ; they have to dry themselves, bathe, bask, cool off, get warm, scrub, wash, wallow, *etc.* One might say that the more highly differentiated the skin, the more it needs attention. Fish, amphibia, and reptiles need not devote such a great part of the day to their slimy or scaly coverings as do feathered or furred creatures.

Here we may draw attention to a point of interest to the animal psychologist, namely, the rate at which many mammals make scratching movements with their hind feet. If for instance a mouse feels an itching on its flank, and scratches the spot with either of its hind paws, the movements that it makes reach such a speed that the human eye is not quick enough to distinguish separate strokes, but sees only a more or less indistinct fan-shaped blur made by the movements of the leg. In larger or slower-moving animals, a St. Bernard dog for example, a beaver, a giant kangaroo, a slow loris, or a sloth, the movements of the foot can be seen while scratching, and the strokes even counted comfortably.

Scratching tempo is thus not only a function of the length of an animal's leg or its size, but of its general rate of movement as well, and so probably also of its " moment ", according to the investigations of several of Uexküll's pupils, particularly G. A. Brecher's (1932). In dealing with animals in the zoo, it is very important for the keeper to adapt himself to their subjective sense of time, *i.e.* to the specific length of their " moment ". Slow-moving animals are scared, sometimes permanently upset, by sudden movements. On the other hand, we, as human beings, consider as threatening not only those animals that react with superhuman strength, but also with superhuman speed ; *e.g.* snakes, which often move as fast as lightning, and to which man, with his longer " moment ", is unable to react properly. Incidentally, we have reached a discussion on subjective " moment " through thinking about care of the skin, to which we must return, keeping within the framework of analysing the character of the animal's daily life.

From time to time, daily life may be completely upset through certain critical conditions of the skin. Every time the crab casts its shell, every time the tortoise begins a phase of growth, and in the case of birds such as penguins, at every moult, there is a real crisis. This usually passes off in hiding, in the greatest possible privacy, as the animal has more or less to suspend most of its other activities, and is greatly handicapped, as far as avoidance of enemies is concerned.

It must not be imagined that the six types of activity mentioned are sharply differentiated. If they could be shown diagrammatically, as sections of a circle—the search for food, the upkeep of the home, social activity, reproduction, care of the body and play, plus two large sectors for sleep and dozing—one sector would enormously increase at certain times, filling nearly all the area of the circle, while all the rest would shrink to a narrow strip. This happens with some animals during rut, for instance. For several days and nights, the rutting stag thinks neither of food nor rest, but uninterruptedly circles the females, in great excitement. The same sort of thing happens with many carnivores, as well as many others.

Thus a universally applicable mathematical formula, with various activities appearing as tidy percentages, is not possible in the case of free-living wild animals. It is, however, possible to draw up average daily programmes for certain animals, and at definite seasons. I should like to choose two examples only, both founded on a very large number of individual observations.

J. M. Linsdale (1946) discovered the following facts about the Californian ground squirrel (*Citellus beecheyi*). He chose a young animal, and a summer's day of fourteen hours. At sunrise, the ground squirrel left its burrow, looked around for a few minutes, defaecated and started looking for food. Then came a long interval for play with other young members of its species. A lot of time was taken up sitting on big stones or posts. Ground squirrels are most often met in this situation. Perhaps we may regard this sitting about as sunbathing, since it only happens in fine weather ; these little rodents, that spend the greatest part of their lives in dark holes underground, have great need of sunshine, but during the most intensive period of sunshine, about midday, the ground squirrel again retired to its shady burrow. From 1 p.m. to 5 p.m., not much activity was to be seen. Scratching and grooming its coat took up considerable time. Later in the afternoon, the search for food was continued, somewhat further away from the burrow, followed by a short play time, until it was ready to go to sleep. According to the weather, ground squirrels stay outside for a longer or shorter time ; on the average, they only spend from one to three hours a day outside their burrows, even in July. This is really surprisingly little. Yet the daily lives of many other animals that have their homes underground appear very similar.

The first-rate American field biologist C. R. Carpenter (1940) gave an accurate picture of the natural daily life of the gibbon, the slender anthropoid ape which swings from tree to tree, based on a painstaking one-hundred-day survey in the virgin forests of Siam. It goes something like this :

5.30—6.00. Reveille
6.00—7.30 or 8.00. Morning singing and general activities
8.00—8.30 or 9.00. Off to the feeding place
9.00—11.00 (approx.) Feeding
11.00—11.30 or so. Walk to a special spot for midday rest
11.30—15.00 (approx.) Siesta ; restricted activity, especially among the young (play)
15.00—17.00 (approx.) More feeding, with change of location
17.00—18.00. Back to sleeping quarters
18.00—to sunset. Preparations for sleep
Soon after sunset until sunrise. Sleep.

It is obvious that all records of this kind, provided they are based on serious observations, are not only interesting from the comparative psychology point of view, but also useful for ensuring a healthy daily life for animals in zoos, where we try to offer them a suitable substitute for life in freedom, as we have already said. The problem of occupational therapy, of providing pastimes in the zoo, if one may express it like that, has recently been given more and more attention, to the great benefit of the animals. We are no longer content to let animals vegetate stupidly in narrow cages, as in the old-fashioned menageries, but do our utmost to see that their lives are healthy and full, and as positive as possible. Yet the indispensable basis for really useful measures such as these remains in the accurate observation of natural daily life in freedom.

4

THE ANIMAL AND ITS ENEMIES

THE MOST important biological objective of all animals, for that matter of all living organisms, and to which all their bodily functions and behaviour are directed, is without any doubt the preservation of the species. Two primary urges serve towards the attainment of this end. Thanks to the poet Friedrich Schiller they have become extremely well known to all German-speaking peoples. In his satire *Die Weltweisen* the poet extols the triumphant ascendancy of nature :

" Einstweilen bis den Bau der Welt
Philosophie zusammenhält
Erhält sie das Getriebe
Durch Hunger und durch Liebe."

" For the present, and until philosophy
Runs the world,
Its driving force it gains
From hunger and from love."

This poetic conception of the maintenance in nature of living creatures has strangely enough become almost a dogma in biology. Even the study of behaviour and animal psychology have been so bedazzled by the predominance of these two mighty urges that for long enough they have almost completely ignored another superior element in the struggle for life throughout the animal kingdom, namely, the all-important escape tendency, *i.e.* the urge to avoid enemies continuously.

"Hunger and Love " take only the second place. The satisfaction of hunger and sexual appetite can be postponed ; not so escape from a dangerous enemy, and all animals, even the biggest and fiercest, have enemies. As far as the higher animals are concerned, escape must thus at any rate be considered as the most important behaviour biologically. The primary duty of the individual, to ensure its own existence, and thus the preservation of its kind, lies in being prepared to escape. By far the chief occupation of the free wild animal, therefore, is constant watchfulness ; eternal alertness for the purpose of avoiding enemies.

There are many examples of species that have died out because their urge to escape, or better still, their flight tendency, like a diseased organ, no longer met the demands made on it. They fell an easy prey to their enemies, *i.e.* to carnivores and man. We only need to recall the giant sea cow (*Rhytina stelleri*), discovered by Steller and killed by hand in great numbers by sailors, whom the animal no more tried to avoid than the sea elephant (*Macrorhinus*) and its relatives avoid the seal hunters of today. The great auk (*Plautus impennis*) and the dodo (*Didus*) allowed themselves to be exterminated because they lacked the urge to escape from man.

An animal that hopes to keep alive among the dangers of freedom must be constantly on the alert. Anyone who takes enough trouble to watch any living creature in the wild state carefully, whether it be a roe deer or a fox, even a mouse or a crow, will at once realize this fact. It is extremely hard to get near them, simply because all animals are so busy keeping an eye open for the possible approach of enemies. As soon as one gets too close, as often happens, they take to flight.

With the exception of a few parts of the globe which have subsequently become secondarily free from enemies, such as remote islands, real safety—rest without danger or undisturbed play, *etc.*—does not in fact exist anywhere for the free-ranging wild animals. All these kinds of peaceful existence occur only in man's imagination, not in nature, although such conditions are usually considered typical of life in freedom.

In actual fact, enemies may be lying in wait anywhere or everywhere. This imposes constant watchfulness on the animal and continual readiness to escape. Even while the animal is cleaning itself, sunbathing, feeding and so on, that is, during the so-called " comfort behaviour " which seems to be typified by a certain relaxation of tension, an appreciable amount of the animal's attention is concerned with safety—with readiness for escape ; *i.e.* with avoidance of enemies. Even during sleep, as we have mentioned, the animal cannot find relief from that most important of all functional spheres of action ; escape from enemies.

In many animals the sleeping position typical of its species is evidence of perpetual alertness. If sleeping animals were not constantly on guard against dangers, *i.e.* prepared for their enemies or ready to escape, hunting and even catching animals would be mere child's play.

I know how hard it is to realize the importance of constantly threatening danger for the animal's behaviour, but anyone who has lived among really primitive natives, as I did on many South Sea Islands, will to some extent appreciate this. These natives felt they were continually being threatened not so much by wild animals but, in an analogous manner, by countless demons. In everything they did, care had to be taken, literally step by step, to ward off evil spirits. Neither eating nor spitting, talking nor sitting, dancing nor hunting could take place without the appropriate measures for avoiding the enemy, *i.e.* the demons.

But let us return to this irresistible impulse towards the continual avoidance of enemies, this flight tendency dominant in the animal's behaviour, and its manifestation in specific flight reaction. This is released by enemies, *i.e.* by animals of other species to which the threatened animal stands in the predator-prey relationship. Man often plays the part of the predatory animal, in fact there is hardly a species of animal that has not been hunted by him, often for centuries or even thousands of years. Thus it may be said that man, with his world-wide distribution and his long-distance weapons represents the arch enemy standing, so to speak, at the flash-point of the escape reactions of animals.

Yet not every approach of the enemy touches off the flight reaction, nor is every approach necessarily a threat. The situation only becomes dangerous when the enemy approaches to within a certain distance of the animal—the escape distance. Only when this specific flight distance, which differs for each species, is overstepped by an observed enemy does flight reaction follow ; *i.e.* the animal in a typical manner runs away from it, far enough to put at least its specific escape distance between itself and the enemy once again.

As a rule, small species of animals have a short escape distance, large animals a long one. The wall lizard can be approached to within a couple of yards before it takes to flight, but a crocodile makes off at fifty. The sparrow hops about unconcerned almost under our feet, thus like the mouse, having a very short flight distance, while crows and eagles, deer and chamois for instance have much longer ones. In Africa, with its abundant wild life, I have, with a tape measure in fact, measured hundreds of flight distances, particularly those of the larger animals, and made comprehensive records of them (Hediger, 1951a). All this confirmed that every species of animal has its particular flight distance, varying of course within certain limits. The animal adapts its flight behaviour to the specific circumstances of its surroundings. Thus, for example, in the national reserves, where man does not appear as a hunter, flight distances are usually less than in the hunting areas. In the Albert National Park the flight distance of the Kob-antelope is reduced to about twenty yards, that of the buffalo, in all other respects so awe-inspiring, to about fifteen yards, and that of the elephant to fifty or so. By intensive treatment, *i.e.* by means of intimate and skilful handling of the wild animals, their flight distance can be made to disappear altogether, so that eventually such animals allow themselves to be touched. This artificial removal of the flight distance between animals and man is the result of the process of taming, defined in animal psychology as the disappearance of flight tendency in the presence of man.

It is important here to show how the animal's flight obeys definite quantitative rules, that the flight distance is of measurable length, and that it can reveal all kinds of things about an animal. But in the qualitative respect also, the animal's flight distance follows precise laws. The animal never just runs or flies away from an enemy. The escape from the threat of a predator or of man obeys strict rules, intimately related to the physical structure of the animal concerned. There is an abundance of actual flight organs that help in avoiding the enemy.

It is a familiar fact that every species of animal has its own special escape technique, its individual flight reaction. The crab swims back, the pheasant flies up, the grebe dives, the mouse makes for the nearest cover, the squirrel swarms up the tree, the hare squats or dashes off, doubling back, *etc.* The example of the hare, squatting in the furrow, or on its form, shows that not every flight reaction necessarily entails a change of place. Escape by suddenly changing locality is the commonest and most obvious way to avoid enemies. There are other ways, like squatting, freezing, shamming death, and, in extreme cases, actual rigidity caused by fright, the so-called " akinesis ", which in a number of cases, including many insects, or the opossum, represents a sort of exaggerated inhibited flight reaction (see the next chapter).

Sometimes a species of animal can be identified far more easily by its particular escape technique, its specific type of flight reaction, than by its external appearance. Thus in New Britain, I often used to come across parties of small lizards (geckos) the colour of tree bark, crawling about walls or fences. It struck me that every time I overstepped their flight distance, they disappeared into crevices, some running upwards, others downwards. This flight reaction in two opposite directions occurred so regularly that I went into the matter. It turned out that this society of geckos consisted of two species, outwardly very much alike, one of which however always ran away upwards, the other always downwards (Hediger, 1934).

Often enough animal trappers have made use of the specific nature of flight reaction, as well as zoologists, who have turned it to good account with vampire bats in Central America. These blood-licking bats were discovered, along with other species of bats, in large caves, and were extremely hard to distinguish. However, their flight reaction picked them out at once, and so the trapper set off the flight reaction of the flocks of bats in the caves illuminated by his lamp, and they rose in great clouds from the walls. The only ones left behind were the desired vampires (*Desmodus*), for their particular type of escape reaction consisted not of flying away, but crawling away, like rodents, and hiding in crannies in the walls of rock where they could be easily captured (Ditmars and Greenhall, 1936).

Even large animals may often have basically different flight reactions in spite of considerable physical affinity. Thus for example, when surprised by its enemy on land the great hippopotamus always makes for the nearest water, while its relative the pigmy hippopotamus (*Choeropsis*) never does so in a similar situation, but disappears into the nearest thicket.

The study of flight reactions and their connection with the physical structure and total behaviour of an animal is of absorbing interest to the biologist. A sympathetic harmony between physical structure, behaviour and surroundings is evident again and again. I should like to give one striking illustration. Tortoises are notoriously well armoured, thanks to one layer of shell and another of bone. This makes these ancient reptiles particularly cumbrous, for armour and mobility are definitely inter-related. On the approach of an enemy, therefore, the tortoise never thinks of running away, but withdraws into the protection of its shell, which in many species can be almost entirely closed thanks to hinged plates. So the predator has little chance of cracking the hard shell open. Among the various species of tortoise, however, there is one exception of interest, the East African soft-shelled land-tortoise (*Malacocherus tornieri*). In this, the armour plating is almost lacking and so flimsy that it moves with every breath. This is known to be a case of secondary reversion of the shell (R. Mertens, 1942). Thus, but for the fact that it alone of all species of tortoise is compensated for by an astonishing flight reaction, the poor tortoise would be defenceless in the face of its enemies. This animal can in fact run away with amazing speed, and reach safety in crevices of the rocks among which it lives. This is almost a classical example of the biological trinity ; surroundings—physical structure—behaviour. Finally let us consider a very different group of armour plated animals, the armadillos. Their behaviour in their natural biotope has been fully investigated by H. Krieg (1929). Different species have plates of different strength, with correspondingly different flight behaviour. Three morphological and ecological types can be distinguished, each associated with special tactics for enemy avoidance and special flight reaction.

The first type, represented by *Dasypus*, contains species with flattened dorsal plates and without ventral protection. On the approach of enemies, these animals dig themselves into the ground or slip into nearby holes with astonishing speed. Representatives of the second type, belonging to the genus *Tatus*, have narrow, highly arched shells and are good runners. Their flight reaction consists of dashing for the nearest cover.

Finally the third type, *Tolypeutes*, has the power of rolling itself up in its shell into a perfect ball. The type most exposed to danger we know from experience to be *Tolypeutes*, the one with the most complete armour, since rolling up into a ball is no use against man, its worst enemy—quite the contrary, it makes capture easier beyond comparison. These armadillos that roll up are favourite delicacies of the Indians, and are killed by them in large numbers. Thus the highly armoured type faces extinction, while the lightly protected *Dasypus* is developing on the other hand into a marked follower of cultivation. As a burrowing animal, *Dasypus* prefers soil that has been broken up by artificial cultivation to the hard ground of the grasslands, and is clever at avoiding the attacks of its enemies by quickly digging itself in. Armadillos that had far stronger armour plating and were much more unwieldy than *Tolypeutes*, e.g. *Glyptodon*, are familiar to palaeontologists, and this genus has a completely rigid set of dorsal and ventral plates. Significantly enough, these clumsy creatures, lacking a suitable flight reaction, died out ages ago. Let us add that, generally speaking, hard plate-armour has been abandoned for greater speed in the course of development of a species.

Up till now we have learnt that enemy avoidance, *i.e.* the animal's flight reaction, is remarkably finely adjusted in quality and quantity. The same thing is true with regard to intensity. If an enemy, recognizable at a distance, approaches slowly, the animal that is potentially its prey withdraws, not in any great hurry but quietly, as the situation demands, and only far enough to preserve the flight distance. I was often able to confirm the familiar fact that antelopes regard fully-fed lions as hardly worth a second glance. They see that the beast of prey is not dangerous this time, and behave accordingly. When, however, an enemy bursts in on them, flight reaction goes off with a bang, so to speak, and it is a long time before they settle down to normal again. I once watched such an unforgettable instance of this, that even today I can vividly recall every detail that I then entered in my diary, although it was several years ago.

It was on the frontier between the Belgian Congo and Tanganyika. After a rewarding day, I was strolling on my own an hour before sunset on the top of a flat hill covered with sparse grass and interspersed with scattered boulders. I sat down on one of these, watching a dozen graceful reed bucks that moved over the area, grazing singly or in small groups, or lying peacefully on the ground. I seemed to be sitting in one of the boxes at some wonderful animal theatre watching an ever-changing and enchanting scene. A bustard was calling upstage as it danced to and fro for a couple of steps, a shapeless ball of ruffled feathers. Peewits flapped up and down, butterfly-like, sometimes tilting at the puffed-up bustard, and startling it. Small oribi antelopes playfully chased each other at the side of the stage, from which four black wart-hogs suddenly appeared, only to disappear again, except for the huge boar that slowly idled up browsing all the time on the bright green grass. On the horizon, a picturesque group of four zebras was silhouetted.

Suddenly the oribis let out a piercing whistle ; the reed buck started up ; the tails of the wart-hogs, until now hanging limply down, stuck up stiffly. All the animals, as though under a spell, looked in the same direction beyond the hill, so that I too looked involuntarily in that direction. Then a reed buck made straight for me, stopped suddenly, and looked round. The situation was one of well-nigh unbearable tension to me. Next a gentle clattering of hoofs became audible. The noise soon grew nearer. What was about to make its appearance upon this stage ? Suddenly a black harnessed antelope buck darted panic-stricken straight across the flat hill-top so close to me that I could see its protruding eyeballs and the foam flecking its open mouth. The other animals watched it terrified, but then quietened down again remarkably soon. Shortly after this mysterious interlude, I got up off my rock and with my binoculars scanned the area from which the excited beast had apparently come, but could make nothing out for certain.

From the evidence, it was clear that the antelope must have had a terrific shock, which had been followed by exaggerated panic flight reaction. I wondered if it had possibly come across Cape hunting dogs, the worst of all predatory animals. A few days later, we did in fact meet a pack of these fearful robbers, beside which lions, leopards and hyaenas seem almost genial.

In contrast to the great cats and hyaenas, Cape hunting dogs (*Lycaon*) have their home everywhere and nowhere. They scour great areas of the African continent in packs, bursting suddenly in on the landscape. There is hardly any effective escape from them. The pack harasses the rear of a herd of antelope in organized detachments, picking out a couple of individuals, and from that moment they are irretrievably lost. After a short chase they are torn to pieces by these speckled robbers, and in next to no time their bones are picked clean. After a bit, a sort of vacuum is created around the bloodthirsty pack; every wild creature gives them a wide berth so the dogs have to find a completely fresh hunting territory, upon which they suddenly fall, only to disappear just as suddenly once more.

Characteristically enough, this super beast of prey has practically no flight tendency in man's presence. The Cape hunting dogs allowed us to approach to within seven yards of them on the open roads, lying around and even playing with a piece of wood that one of them had picked up. Hunters and game wardens have told me how much they dislike having to shoot these wild dogs, since they approach unbelievably close to the guns, showing no reaction to the shooting of their fellows so that they can be picked off at ease, one after another. It often happens too that the first to be shot are at once fallen on by the survivors and torn to pieces like prey as soon as the latter smell their blood.

In Africa it has frequently been observed that antelopes, in panic flight from Cape hunting dogs, dash through the lines of native carriers and even right into camps or villages, where they seek safety from their worst enemies, trembling all over and completely exhausted. In the Far North too, similar reports have been given about reindeer, which sought safety from wolves even among the ranks of the stalwart musk ox that were taking up their typical defence formation.

No doubt the Cape hunting dog presents a rare and extreme case. In most cases predators and prey live as it were side by side, *i.e.* their territories, even their homes overlap. Antelopes and zebras are more or less reconciled to the presence of lions on their feeding grounds, which are a feature of their living space: fate personified, so to speak. As long as the predator is not hungry, not much notice is taken of it, and the sudden sharp shock of the kill is comparatively soon forgotten by the survivors. Constant other threats mean continuous activity, and necessitate attentiveness in the service of enemy avoidance, until at last alertness fails.

It is vital for the predator to be able to kill enough prey ; for the latter it is critical to be able to escape from the enemy. Almost every feature of bodily structure, coloration, especially external appearance of the victim, is aimed at helping to avoid enemies, and to make escape as successful as possible. Frequently the whole structure of the victim is, if I may so express it, tuned to the characteristics of its chief enemy, or enemies; so that, for example, in the case of the common hare, one may almost speak of the enemy specificity of its escape behaviour or flight reaction. The hare has two quite separate escape reactions; one against attack from the air, the other against land enemies. The tactics for the former, especially for the hawk, consist of squatting motionless on the ground. For terrestrial enemies, such as foxes, hounds, *etc.*, it relies on running away at lightning speed, with characteristic doubling on its tracks. Many birds as well, such as the house sparrow, plover and jungle fowl have two specific flight reactions to enemies, with and without movement away from the spot, corresponding to their two main sorts of enemies, air and ground enemies.

In connection with dual flight reaction, there are normally two different warning signals present, depending on the presence of a threat from a bird of prey, or from a land animal. Here we enter the field of escape signals, by which the presence of the enemy is registered, and preparation for escape made. We must stress the fact that, contrary to popular opinion, the animal giving the signal has no intention of warning others, though such warning may in effect result. Here again, we may distinguish between two groups of escape signals: 1. Intra-specific; *i.e.* those which serve to inform members of its own species (warning); 2. Interspecific; which warn members of other species as well. Thus for instance the shrill scream of the wild peacock sets the whole jungle alert, while the piercing whistle of the reed buck causes other antelopes and even ibis and heron, to prepare for flight.

Escape signals can be of three types: acoustic, optic and olfactory. Acoustic signals are the most familiar. Most Central Europeans, for example, know the characteristic whistle of the chamois and marmot, or the jay's warning screech. The rabbit drums its flight signal surprisingly loudly with its hind feet, and the frog makes use of the so-called splash sound, flopping his body on to the water surface, and so warning his neighbours. The beaver splashes his trowel-like tail loudly on the water. Even some fish, *e.g.* minnows, give audible flight signals.

Optical escape signals are found among certain *Cichlids*. Among many social birds or mammals, the flight of a member of its species is contagious, so to speak.

Frequently conspicuous patterns or colours are displayed on the hind quarters, acting as flight releasers and setting off escape reaction among their fellow animals. A wild goose makes a striking transformation while in flight by suddenly showing its white back with the contrasting tail; similarly, changes occur during flight in the sika deer, reindeer, or antelope through the expansion of the white patch on the rump, or the stiff erection of the tail.

Finally, olfactory signals may put to flight other members of the species. K. von Frisch has pointed out (1938) in fundamental experiments, that minnows detect fluid extracted by crushing from the skin of other members of their own species, and respond at once with flight reaction. At the actual spot where a frightened reindeer has left secretion from its interdigital glands, all the reindeer that follow take fright.

The connections between escape and soma (physical structure) are very diversified. Let us take external appearance first. Any over-conspicuous feature runs counter to the basic necessity for avoiding enemies. Thus we usually find nothing in the animal to make it stand out against its background, but, on the contrary, the greatest possible adaptation and this visual adaptation may lead to complete identification with it. Extreme cases of this phenomenon used to be described by the term " mimicry "; we now talk of somatolysis, visual adaptation, *etc*. The typical wild animal's coloration is the so-called wild colour, and is as inconspicuous as possible.

Heinroth (1938) thinks that song has developed to such a wonderful degree in birds only because, as fliers, they have a far greater capacity for escape than earthbound mammals. Here again, analogous to optical conspicuousness, the use of the voice is still further intensified under domestication. Conspicuousness of vocal expression may be increased not only by domestication but also by captivity and taming, *i.e.*, by securing from enemies not only individuals, but generations of animals. Imitative parrots, for example, are unknown in nature.

It is obvious that the receptors, the sense organs, must be properly tuned to recognize enemies, especially the chief enemy, so that escape can be assured at the right moment. In his *Bedeutungslehre*, J. von Uexküll (1940) quotes the excellent example of a night moth that is deaf to all sounds except one, namely, the cry of its arch-enemy, the chirping of the bat. The moth's hearing organs consist of two tightly stretched processes that act as resonators. These are tuned to air vibrations which, to man's ear, lie on the threshold of audibility, and exactly respond to the delicate shrilling of the bat. According to Uexküll, only sounds coming from its chief enemy are received by these moths; otherwise, to it the world is a silent place.

G. Billard and P. Dodel (1922) have distinguished two groups, hunters and hunted, among the higher animals in connection with the development of receptors. These terms are somewhat misleading, because all animals are literally hunted in freedom, even the most obvious hunting animals. Every animal has enemies; herbivores must keep constant watch for carnivores, lesser carnivores for greater, and all animals for man.

With hunters, that is, carnivorous mammals and raptorial birds, the eyes are set frontally, the field of vision lies directly ahead of the animal, and is binocular; the visual axes intersect; the jaw muscles behind the eye are powerfully developed.

On the other hand, the hunted, the herbivores, have often very prominent eyes set at the sides; the field of vision extends to the sides and to the rear, and is unrestricted by the relatively weak and much less prominent jaw muscles. The visual axes do not intersect; each eye has a separate field of vision. Ever prepared for escape, the hunted can thus scan almost the whole horizon. Only with difficulty can an enemy approach.

In addition to the particular set of the eyes in either type, there is also the special shape of the pupil. The hunted often have broadly oval, wide-set horizontal pupils, that always remain horizontal independently of the most violent movement of the head, *e.g.* in Equidae and ruminants (K. M. Schneider, 1930). In contrast to these, the hunters have circular or vertically slitted pupils.

Apart from the organs of sense, or receptors, the organs of locomotion, or effectors, are especially linked with flight behaviour. Most birds rely on their powers of flying in order to escape. During moult, therefore, they must not lose all their pinion feathers at the same time. Only a few waterfowl, such as ducks and geese, can afford to do this, as they do not entirely rely on flying in order to escape, but can paddle away or even dive.

The Australian semi-palmated goose (*Anseranas semipalmata*), that practically never enters the water, is an exception ; its feet are not typically webbed, but, as the name implies, have practically no webs at all. This goose, emancipated from the water, finds it impossible to save itself in an emergency by swimming or diving; it must rely on flight to escape. It is of significance that its method of moulting lies quite outside the range of its related group. This is the only goose which does not lose all its pinion feathers at once, but replaces them singly, so that its powers of flight remain unimpaired.

There are organs of locomotion exclusively used for escape, and they are always the most efficient, while other, slower-working ones serve for normal locomotion. The fresh-water crayfish, for instance, walks forwards on its spindly legs, but, for the purpose of escape, and then only, it brings its strong abdominal muscle into play, and swims backwards. It would be justifiable to call the musculus adductor abdominis its flight muscle.

Or let us consider the octopus, that draws itself forward by its sucker-covered arms. When it has to escape, however, it squirts a jet of water, with the help of its strong mantle muscle, with such force through its respiratory tube, that it is propelled backwards in a flash by the reaction. The earthworm generally employs the characteristic method of peristaltic locomotion. When it is being chased by its chief enemy, the mole, which causes underground vibrations to which it is very sensitive, a completely different method of locomotion is brought into play, namely, vertical undulation, only to be seen when it is escaping.

Nearly all flying birds have adopted special tactics to ward off attacks from above. According to Lorenz (1940) this is a special kind of instinctive action. It consists of a swift roll on to the back on the part of the bird under attack, and consequently a sudden downward dive. The bird being chased then doubles back as it were and flies straight on again.

Pigeons also double back when escaping from birds of prey. By means of breeding, man has now developed a race of pigeons, the tumblers, in which this specific behaviour has become, as it were, chronic. Normal methods of locomotion have been artificially bred out of them, and all that is left is the locomotion of escape. Normal and escape locomotion can be so different that they are separately influenced by breeding methods.

If the close connections between escape behaviour on the one hand, and the structure of receptors and effectors on the other are strikingly obvious, the connections between escape behaviour and the special structure of internal organs are naturally not so easy to observe. Yet a critical inspection will hardly discover any system of organs that has remained unaffected, in individual cases, by the need for escaping. We need only mention here Beninde's hypothesis (1937) on the development of the ruminant stomach in the Cervidae.

The deer's differentiated stomach no doubt serves today for the better assimilation of a diet lacking in nutritive value. Beninde, however, thinks that we must consider irresistible escape impulses as the formative agent. The decisive factor in the "invention" of the ruminant stomach, says Beninde, speaking of deer, was evidently the attempt to reduce to a minimum the dangerous time needed for feeding in open country. A large quantity of food is taken at great speed into the rumen, without time being wasted on mastication. Reduction only takes place later under cover, in the form of chewing the cud.

There is much to be said for the hypothesis that this hasty snatching of food in open country, with exposure to enemies, followed by comparatively leisurely mastication in relative safety under cover, may well have found organic expression, since parallel phenomena have been observed in the most widely differing groups of animals. We need only mention the crops of birds, or the cheek pouches of many rodents.

Having thus considered the multiple relationships between escape behaviour and body organs, and quoted a few examples, let us turn for a moment to the hypothetical development of this important category of behaviour. The Dutch physiologist, J. ten Cate (1938), has studied localized reflexes in *Branchiostoma*, that form of living organism which lies at the root of the great and many-branched tree of the vertebrates. It appears that localized reflex actions are absent in the whole locomotor system of this primitive creature. Whatever portion of the *Branchiostoma's* body was stimulated, says ten Cate, an undulating movement always followed if the stimulus was effective, and this spread throughout the whole body, leading finally to escape reaction.

In *Branchiostoma*, the invariable response to all disturbing stimuli from the external world is total reaction of the whole organism, *i.e.* an undulatory movement of all the body, identical with the animal's escape reaction. This undulation may be described as the basic form of escape, the original escape reaction, of all vertebrates. It also occurs in other primitive vertebrata, *e.g.* the *Cyclostoma*. Characteristically, it appears also in the embryonic stages of nearly all vertebrates, as Coghill and others have shown. All

Figure 6a.—Female Himalayan tahr (Hemitragus jemlahicus), *after a spontaneous visit to its next-door neighbours, three male Barbary sheep* (Ammotragus lervia), *jumping over the six-foot gate back into its own quarters, where its young is. Example of unusually peaceful relationship between animals of different species. The leap takes place from a standing position, without any run-up.*

Figure 6b.—The female tahr jumping over.

Figure 7.—*Blackheaded gulls* (Larus ridibundus) *on a floating boom at Oberwil on the Lake of Zug (22 July, 1950). The specific individual distance is conspicuous.*

embryos share a fluid medium, and an oblong shape, with *Branchiostoma*. According to this, it seems therefore that the whole complement of individual reactions of the vertebrate animal is to be attributed, phylogenetically and ontogenetically, to differentiation of the basic escape reaction.

Finally, mention must be made of the wealth and diversity for man of the significance of escape in animals. Here a practical and a theoretical side are distinguishable. Analysis of the behaviour of domestic animals shows how intimately these can affect each other. B. Klatt (1937) was justified in referring to the creation of domestic animals by man as " the oldest and, in its scope, most grandiose example of man's activity in experimental biology."

Yet this gigantic experiment was not carried out as the result of one particular investigation, or by special individuals. It started in complete ignorance and anonymity, and continued for many generations. The result is that today we are confronted with the result of this experiment, the highly trained domestic animal, without being able to understand it essentially. It is certain, however, that all domestic animals have come from wild ancestors, and represent the material basis for the origin of human culture.

If we look more closely, it becomes clear that the primary reason for usefulness as domestic animals to man is not the horse's pulling powers, nor the dog's intelligence, nor the cow's milk capacity, nor the hen's egg output, but is basically a very different domestic quality, namely, the disappearance of that tendency to escape, so fundamentally important for their wild ancestors. The most active horse or the most productive milch cow would be of no use to us if these animals still possessed in undiminished form the escape tendency of their ancestral wild types. When the original flight tendency asserts itself, if only for a few moments, as in the case of a stampeding horse, there is a catastrophe (*see* Chapter 8). Escape tendency excludes usefulness. The proof of this lies in those domestic animals which have gone wild.

Removal of escape tendency, *i.e.* taming and tameness, can only come from man. He is the only creature capable of freeing another from the magic circle of flight, from the irresistible impulse to avoid enemies continually. Man is moreover the only creature able to free himself from the elementary function of escape. By this self-release, man clearly stands apart from the rest of creation, and, as the arch-enemy, is the focus of all animal escape reactions.

Release from the bonds of the flight circle, either in animal or man, implies at the same time the freeing of powerful energies, all those in fact which had previously been devoted to the impulse to escape from enemies, the constant tension in order to be ready for flight. If one considers once again how the whole behaviour of the animal is dominated by the necessity, the overriding necessity for escape, and when one realizes that this all-powerful factor of flight is being wasted, one involuntarily wonders how this vacuum is to be filled. In the tame animal under experiment, released energy appears in the form of an astonishing mental efficiency, compared with the animal in freedom. The creativeness of the energy released in man makes its appearance in the beginnings and development of culture, which has reached different levels in different races. Among primitive natives we find, as we have said, that they have broken out of the animal's escape circle, but are none the less at the mercy of the still irresistible urge to escape from the countless demons that infest their subjective world. Thus major importance must be attached to release from the escape circle.

The connections existing between the original flight tendency and the beginnings of culture have not escaped the historians and philosophers either. Many of them see in ancient architecture, especially the ecclesiastical architecture of the early Middle Ages, in the carvings of demons, dragons' heads and animal masks, a clear indication of the fear of devils in the men of those days. They consider the Gothic style of architecture as the deliverance, the release, from the gloom of the flight circle, from the fear of demons, which still kept the man of the Romanesque period, and even more primitive man, in chains.

In bringing this chapter to a close, we may as well recapitulate the situation of the free-living wild animal, characterized by incessant threats, and the never-ending necessity for avoiding enemies. In the zoo, where the animal's existence is protected by bars against external enemies and where its "field tension", to quote G. Bally (1945), is to a large extent relaxed, it may enjoy a kind of rest that is quite foreign to it in "golden freedom". Anyone who watches animals in the zoo closely must receive the impression that they are enjoying this positive side of existence inside bars. This absence of compulsory activity must even be compensated for artificially as much as possible, and on this we shall have something to say in a later chapter.

5

" FLIGHT " AND " HYPNOSIS "

Now that we have seen in the previous chapter how dominant escape behaviour is, and realized what a fundamental part it can play in forming the external appearance of the animal, as well as its mental make-up and the development of the internal organs, it will come as no surprise that many kinds of behaviour, which are apparently only remotely connected, can clearly be linked with escape behaviour.

Among those special types of behaviour which may be considered as particular cases of escape behaviour, are phenomena called—not very aptly —" animal hypnosis ", or akinesis; " action inhibition ", " thanatosis ", " shamming dead ", are other terms that have been used in this connection. The question with which we are concerned now is one of enormous complexity, and is already fully documented. To deal with it in this short chapter would be presumptuous. Anyone who would care to pursue it in greater detail is recommended to read F. Steiniger's excellent study (1936). Here we say no more than to point out that certain forms of loss of movement, hitherto considered by some authors as belonging to the category of " animal hypnosis ", may now be regarded as special cases of escape behaviour. In addition I should like to show that certain kinds of behaviour between carnivores and their victims, especially snakes and their prey, so often the subject of questions in the zoo, are connected with escape behaviour.

The primitive impulse of every healthy creature to preserve life is the starting point. This becomes evident mostly through threats by enemies, from which it is vital to run away in time. Enemy-avoidance usually consists of sudden and effective change of place, that is, of the specific escape reaction. This sudden change of place is not always adequate, some species, e.g., hares, bitterns and many others keep still, and in extreme cases rigidity, or shamming death, sometimes seem more effective.

In many animals, in which this contrary kind of behaviour in the form of escape reactions specific to the enemy do not normally occur, it may sometimes be induced. The Swiss psychiatrist L. Greppin (1911), who has published many valuable studies of animal psychology, came across a classic case of this type. As the director of the Rosegg sanatorium and nursing home, he was responsible for the extermination of the sparrows that swarmed in the grounds of his institution.

When the kill was on, he soon noticed an increase of from thirty to a hundred or a hundred and fifty yards in the sparrows' escape distance from the men with the guns. The escape distance from unarmed men did not materially alter. When the sparrow-hunt intensified, the following was noticed: " At first, they only discriminated against me when carrying a gun; then, whether armed or not, I was distinguished from the rest, and at the same time, flight distance increased. As soon as I appeared, their cries of fright and alarm grew in number and intensity, and the escape reaction began still farther away. Then, after about eight or ten weeks, unusual phenomena of movements and lameness appeared as an expression of almost panic fright. The sparrows dropped like stones into the bushes, and then looked as if they had stiffened into rigid postures."

Thus, thanks to an intensive threat, the sparrows were caught up in a kind of akinesis; lameness had evolved from rapid flying away. In zoo aviaries, too, during cleaning or other disturbances, this may sometimes be seen. Rigidity of this sort may also prove effective against four-footed enemies. C. Heinroth (1933) confirmed this by a chance observation on sparrows. " In Berlin Zoological Gardens, a domestic cat had caught a sparrow, and was carrying it towards the Great Cats' House. Then she started to play with it; she dropped it on the ground, and every time it tried to fly away, knocked it back on to the ground again. This was repeated several times. At last the sparrow kept still for some time, as if it were dead, so that the cat's attention wandered a little. Suddenly the bird flew off to freedom, while all the cat's trouble was for nothing. This was proof to me that the victim's shamming dead had saved its life."

In addition there are classic analogies in those few cases where men have been the victims of the larger carnivores. No less than David Livingstone (1813–1873), the famous missionary and African explorer, himself became the unwilling subject of an experiment of this sort. Livingstone was attacked in remarkable circumstances by a lion which shook him, just as a terrier shakes a rat. The shock caused a sort of stupor or semi narcosis, during which he was able to register every detail of what was happening, but felt neither pain nor fear.

Another explorer, Inverarity, was also seized by a lion, in Somaliland. In contrast to Livingstone, however, he felt pain throughout. Far more painful than the flesh bites was the pressure of the teeth on the periosteum. Both victims kept as still as though dead, and, as Burton says (1931, p. 58), owe their lives to this very fact of absolute motionlessness.

These examples, particularly that of Heinroth and the sparrows, show clearly the positive results of rigidity in helping to avoid enemies. Shamming death in the face of danger is very common in the animal kingdom, and is even commoner among invertebrates than vertebrates. Absolute stillness, as well as quick escape, helps to avoid enemies. F. Steiniger has already drawn attention to this fact (1938), for which there is ample evidence. He mentions how very slender the opportunities are for comparing such " animal hypnosis " with that in human beings. His observations are perfectly in harmony with the conception that such forms of rigidity represent from the biological standpoint nothing more than practical behaviour aimed at escape, thus at enemy avoidance.

This is so not only for various insects, crabs, amphibia, reptiles and birds, but for many mammals as well. The situation of absolute surrender to a superior enemy can most easily be produced experimentally by putting the animal subject on its back, thus repeating the well-known *experimentum mirabile* demonstrated as far back as 1646 by the Jesuit Father Athanasius Kircher, and since then repeated, varied, and discussed, time and time again. This has little to do with human hypnosis, and to that extent the term " animal hypnosis " is quite misleading.

Showmen and a certain class of pseudo-scientific writers have repeatedly chosen this sort of rigid condition as an opportunity for sensational acts and performances, often with surprising results. Among these, to take one of the worst examples, is Völgyesi's *Human and Animal Hypnosis*, which was quite unnecessarily translated into German in 1938. For an illustration of this decidedly pseudo-scientific kind of work, see the legends to *Figures 14* and *15*. " The Giant Brazilian Snake (*Python mulurus*) hypnotises, then swallows a hare rigid with fear." In the first place, we must point out that no *Python mulurus* is known to science, but only a *Python Molurus*, nor does this occur in Brazil, but in tropical Asia; then the animal in the picture is not a hare at all, but a young rabbit. The description of illustration No. 12 is also all wrong, as is by far the greater part of this fantastically absurd book.

Undeniable cases among various reptiles of shamming death in dangerous situations have been described by Th. Barbour (1934). The monitor lizard of the Congo (*Varanus exanthematicus*), when taken by surprise, may suddenly turn on its back and in this position bite its hind leg. The North American snake Heterodon turns on its back similarly, opens its mouth, and hangs its tongue out. It looks as if it had just been killed, but makes the mistake of immediately reverting to this state if it is put back into its normal position. The same thing happens with the great Central African tree toad (*Bufo superciliaris*).

If one takes into consideration the fact that many predators locate their prey in the first instance by its movements, and are then attracted to it, such behaviour methods are bound to have a life-saving result, and are thus without any doubt aids to enemy avoidance. R. Mertens (1946) has published a detailed monograph on warning and threatening actions of this sort. Another problem is this: how far are such types of behaviour under the control of the animal, and how much have they developed through experience ? In the great majority of cases, it is probably a matter of definite fixed pattern, beyond the animal's control.

This is apparently true even for mammals, among which the most famous example is the American opossum (*Didelphys*). As W. J. Hamilton (1939) describes with illustrations, it drops down, shuts or rolls up its eyes, and lets its tongue loll out of its half-open mouth. This strange attitude may last several minutes, until the danger is over.

We might mention here the experience of many poultry-breeders. Brood hens often escape with their lives, when attacked by martens, through sitting quite still on their nests and not provoking the predators to attack by their attempts to escape. There is no doubt about the extreme importance of keeping still as an aid to defence against enemies.

From the biological point of view, it would seem that in many species of animals, from insects to man, definite modes of behaviour are pre-established, appearing under situations of threat from enemies. In other species, similar conditions of temporary akinesis can be induced by appropriate human intervention, such as by forcing them to adopt a supine position by a sudden shock. This has no direct connection with hypnosis in the human sense. We sometimes use this immobilizing in the zoo ; for example, when weighing small animals.

The second sort of " hypnosis ", which one might call a special case of flight, concerns the fascination of their prey by snakes. B. Bavink (1949, p. 587) speaks of the " well-authenticated ' bewitching ' of birds by snakes." I have nothing fresh to add on this score; on the other hand, G. Hinsche (1928 and 1939) has made very important statements about the relations between snakes and amphibia (frogs and toads), to which we now refer.

It is my intention to show what is meant by this "bewitching". For more than twenty years I had almost daily opportunities of watching any number of different snakes in the zoo, and in freedom as well. Experience has taught me that it would be incorrect simply to assert, as do many over-objective scientists, that such a thing as fascination does not exist. I fully appreciate the reasons for their opinions, since white mice or guinea pigs can be put in front of snakes literally hundreds of times without these food animals ever being anything but completely unaffected by the snakes, or even behaving indifferently towards them. I myself was somewhat anxious about the snakes, when rabbits hopped about over them, as they did over any other kind of floor, and thus scratched them, or when for example white rats gnawed the rattles of the rattlesnakes or even nibbled their scales. Of course, as many enthusiasts for vivaria will testify, similar observations have been made in every zoo where snakes are kept. Yet it would not do to go so far as to assert that there is no such thing as fascination, on the basis of such observation. An essential supposition for the emergence of a relationship of this kind seems to me to lie in using as victims not laboratory animals, but wild ones normally occurring in the snake's biotope. How can a white mouse or guinea pig, living for generations in laboratory cages, react properly to a snake ? Possible fixed patterns of reaction to the snake as enemy must have died out long ago, like the other primary behaviour of wild animals, and no opportunities occur in the laboratory for acquiring learnt modes of behaviour. On the other hand, we may imagine that genuine experienced wild animals have special reactions to snakes, especially when these are the chief enemies in their biotope, just as other animals have to their most important enemies.

Apart from these, however, it is in my opinion justifiable in some cases to talk of fascination, especially in the sense of an arresting of the victim's attention by snakes. Of course, this phenomenon has nothing to do with hypnosis in the human sense, but with escape behaviour, as we shall see. Many snakes have fascinatory organs; devices, that is, ideally adapted for attracting the attention of their quarry, but without causing it to escape. The snake produces sub-normal flight stimuli, calculated to keep the animal in a constant state of tension, without running away. Frequently these literally fascinating (*i.e.* binding) phenomena possess a bait-like character, so that the victim becomes completely puzzled, and has to stay to solve the mystery. This is what gives an observer the impression that it is spellbound.

Two rare opisthoglyphous snakes from Madagascar provide some of the most grotesque examples of this kind of thing: *Langaha nasuta* and *Langaha crista-galli*. The first has a long finger-like process projecting from its snout, the second something that looks like a cock's comb, *i.e.* a tufted or branched appendage. When a figure of this sort slowly appears before a victim, the latter naturally needs time to decide what sort of thing this weird creature is. He is " fascinated " by it.

Quite a number of similar fascinatory organs are found in snakes, and on various parts of the body. The water-snake from Indo-China, *Herpeton tentaculum*, even has two movable nasal appendages, that can be extended and retracted like worms. In *Thelotornis kirtlandii* from Central Africa, the tongue is largely specialized as a fascinatory organ. In this snake, it not only darts flickering out through the rostral aperture when the mouth orifice is closed, as in other snakes, but is vividly coloured, and makes the strangest movements. The snake itself is a tree-dwelling species, and looks remarkably like a liana. When the snake is after a victim, it lies still at a suitable place among the branches and rapidly flickers its brightly-coloured, black-tipped forked tongue. According to Pitman (1938), the tongue then looks like an insect. Birds and lizards become interested in it, and it takes some time for them to realize what the decoy is. Their attention is riveted while the snake itself prepares to strike.

The tongue, normally of an inconspicuous grey or reddish colour, has a violently coloured pattern, and performs the most unusual contortions. For instance, it is protruded from the mouth with the two tips touching; then these proceed to separate until they are almost at an angle of 180°, after which they are brought together again. The tongue, however, may remain for some time stiffly protruding and eventually be waved up and down slowly with the tips outspread (Barbour, 1934, p. 48).

Strangely enough, a very similar fascination mechanism is found in a South American snake (*Oxybelis*), likewise of liana shape. In this species too the tongue can be pushed far out, and deceive a victim (Curran and Kauffeld, 1937, p. 120). Like *Thelotornis*, *Oxybelis* can also puff out a vividly coloured portion of its neck. According to J. B. Procter (1924, p. 1127), the tongue of *Oxybelis* is olive-green on the upper side as far as the tips; the sides, like the whole of the snake's body, have a black horizontal line, whilst the lower side is cream coloured. The snout of this slender tree snake is unusually narrow, so that the protruding tongue looks like a continuation of the head, especially as it can also appear through the rostral aperture when the mouth is closed. This gives the impression that the fore end of the snake is alternately expanding and contracting, and this again attracts the attention of a victim to an extreme degree.

This device serves (as Procter terms it) to fascinate lizards. Miss Joan Beauchamp Procter who was Curator of Reptiles at London Zoo for many years from 1923, had an unusual wealth of experience with these creatures, as well as the good luck to be able to watch this fascination behaviour once in *Oxybelis*. " I have seen it doing this, the lizard watching the hovering tongue in perplexity, quite engrossed, and the snake striking at close range without startling the usually swift and cautious prey."

In other snakes, the fascinatory organ is situated at the rear, not the front. Among certain Viperidae and Crotalidae, the tip of the tail assumes this function. C. H. Pope (1937) has a note on young Crotalidae from Ceylon (*Ancistrodon hypnale*) in the snake pit, to which a small lizard was presented. The tips of their tails, brilliantly coloured on the underside, immediately began to perform weird, worm-like movements. Lizards have repeatedly been seen to bite at this tail decoy, thereby immediately inviting the deadly poison bite of the hungry snake. This deceptive use of the tail also occurs in other snakes with vividly-coloured tail tips, according to Pope; in for instance the Crotalidons, *Bothrops atrox*, and *Agkistrodon mokasen.*

Tail movements in snakes may have basically different meanings. Here we are only concerned with those which have a fascinatory effect on victims. At the same time, as at the head end, there are gradations from true fascinatory mechanism to deceptive lures. The difference lies in the fact that the fascination apparatus only serves to rivet the attention, while the lure, performing the same function, is immediately taken, and is thus regarded by the victim as edible. In this case, the snake's victim is completely deceived; the snake no longer appears as an enemy, but, in part at least, as a prey. Typical fascination is not produced by the obvious deception of a decoy, but by something of indefinite quality that causes the animal in its perplexity, to make a thorough examination.

Either purpose fulfils its function biologically; it delays or hinders the escape of the victim, even turning flight into approach, thus simplifying capture.

In snakes we must distinguish between simple tail movements, and vibrating usually connected with the production of noises, and which may differ widely in rhythm and intensity in various species. This vibration occurs in the genus Spilotes, Zamenis, Coluber, Pituophis, Ophibolus, *etc.*, not in the situation of hunting for a victim, but as an expression of states of excitement, caused for instance by the sudden appearance of enemies. The poisonous water mocassin (*Ancistrodon piscivorus*) vibrates the tip of its tail, showing a form of behaviour that has reached its culmination in the rattle-bearing Crotalidae. In these, another noise apparatus appears. This kind of tail behaviour has no connection with fascination mechanism, but is more of a warning device.

As we have said, fascinatory organs are found especially at the head and tail. Some however lie between. An example of this is Dahl's snake (*Zamenis dahli*), that lives in Dalmatia. This elegant snake has a habit of undulating its neck, while looking for a prey, its head remaining stationary. It is clear that the attention of the intended victim—a lizard maybe—is drawn to this waving movement, so that the immobile head can gradually approach the prey to within striking distance.

Thus in Dahl's snake, it is not the head, but a section of the snake's body, that forms the fascinatory organ. From this to the non-specialized snake, so to speak, is only a step. Basically, every snake's body is a fascinatory organ ; it has no limbs, no ear flaps, not even a distinct division between head, neck, trunk and tail; nor has it movable eyelids or externally visible face muscles, not even a vivid change of colouring, *etc.*

It is here that Hinsche's work, already referred to, gains special significance. This author has described a large number of flight reactions in various native amphibia, hitherto seldom observed. These mostly concern types of behaviour produced in particular situations by outstanding enemies, and which are phylogenetically speaking very old; " stegocephalic " types, as it were.

The close connection between these types of behaviour and escape, or enemy avoidance, always the focus of our attention, follows from the following statement of Hinsche (1939, p. 729): " When a victim is attacked, there are in principle two courses of action open to it; to escape, or to remain. The latter may be divided into several complexes of reaction: akinesis (death shamming), defence, defensive attack, *etc*. To remain is thus effective, provided the opponent is not more powerful, or when no possibility of escape is left.

From our point of view, the interesting behaviour occurs in a situation of dual tendency—to escape or to remain. Anyone who has watched the grass snake when it is ready for food, knows that it does not recognize the frog sitting motionless; but the moment it hops away, the snake darts out and seizes it. There can be no doubt that the victim's immobility often results in its not being discovered by the enemy. As an alternative to escape, active defence may sometimes replace immobility, perhaps through swelling up; through a strange stilt-like rearing up and turning the broad upper surface of the head and trunk outwards; through partly standing on the head ; and even through butting or screeching."

In the Zoological Gardens at Halle, according to Hinsche, a few whistling frogs (*Leptodactylus ocellatus*) were for a long time kept in the same tank as a 12 foot *Python molurus*. The frogs took no notice of the snake, but hopped around unconcerned while the latter, for its part, ignored the amphibia. Once the snake was excitedly searching the tank for a rabbit that it had previously disgorged. A whistling frog happened to be sitting in a corner, from which no escape was possible. As the snake approached it half drew itself up on its legs, puffed itself out and butted at the snake with its head. Although the snake at once made off, the excited frog remained for about three minutes in this strange posture.

This example shows, among other things, that only in very special conditions does behaviour other than escape from (or indifference to) the snake occur. The stilt position, akinesis, rigidity, fascination—whatever we care to call them—are in actual fact seldom observed; yet such states do occur, and are of extreme interest. Precisely similar behaviour, experimentally provoked by Hinsche, has already been accidentally seen by other observers. I recently came across a report by E. E. Green in the Indian scientific journal *Spolia Zeylanica* (1906, p. 196). In it, he describes the behaviour of the widely distributed Indian snake *Tropidonatus stellatus*, which feed exclusively on toads.

Normally this snake takes its prey unawares. If, however, the toad notices its enemy in time, it adopts a curious attitude, clearly puzzling to the snake, which is often even put off by this tactical surprise. The toad squats forward until its mouth touches the ground, at the same time stiffening its legs into a stilt-like posture and puffing itself right out. If the snake does not retreat at this transformation of its prey, the other tries different defence tactics, alternately bending and stretching its legs so as to produce a remarkable rocking to and fro of its body in the direction of the snake. Green is correct in assuming that this must be a case of inherited and not of individually acquired behaviour.

57

Under the stimulus of this account, E. E. Austen published a very similar observation in the same periodical (1907), made by him, not in India but in England, and in a greenhouse. The toad was the common native *Bufo vulgaris* and the snake, a grass snake (*Natrix natrix*). The behaviour of these protagonists was exactly as Green had described it in his account.

The Arabs of North Africa claim that the chameleon puffs itself out irregularly with the help of well developed air sacs on sighting a threatening snake, and holds a piece of wood transversally in its mouth to prevent itself being swallowed. The puffing up may be true, but no zoologist would believe in the stick. This is a case of the human imagination projecting itself in an interesting way along a biological direction, so to speak.

In most cases no reaction similar to the one closely investigated by Hinsche can be observed, as we have mentioned ; they usually occur in special circumstances, mostly when the victim is already in a state of excitement and uncertainty, when escape is impossible, or the opportunity for it already past, and when the snake finds itself in a visually favourable position. By this, we mean the creeping up, or partial emergence, of the snake's limbless body, already described, with its single conspicuous points such as the tongue or the appendages mentioned, or the undulating parts. The effectiveness of the visual appearance is increased according to Hinsche, " when a clearly visible part of the object is adjacent to others less visible ; conditions which are fulfilled when separate parts of the indistinct image of a larger object—the sharp snout of an assailant, perhaps, or the flickering tongue—stand out more sharply and present the specific stimulus."

In a certain sense, as we have said, the whole of the snake's writhing body is difficult to take in at a glance, and, with its front end partly conspicuous, and partly camouflaged, is a striking fascinatory organ. Of particular interest is Hinsche's statement that he succeeded in producing the same reaction in frogs and toads, which were greatly excited and rendered sensitive by previous stimuli, with the help of a dummy snake, an ordinary rubber hose that he coiled and twisted close to them. This shows very well how the mysterious appearance of the snake depends, in this connection, simply on the visual effect of its shape, and not on some mysterious powers.

To many human beings, the snake is not just an animal of occasionally puzzling appearance, but the embodiment of a Something, unique among animals. For them, in short, snakes are in themselves mysterious and repugnant in the highest degree, nor is this connected with their venom. While running a zoo, one can come across examples of this which are of greater psychological interest.

As a schoolboy I had an experience of this kind, having been given the job of fitting up the long-disused tanks in a corridor of the biological department. One half holiday I turned up with a decorative piece of tree root, moss, and the necessary grass snakes. As I was admiring the freshly furnished tank, a cleaner came up with her mop and bucket, and looked at the beautifully fitted-up tank with apparent interest.

The moment she caught sight of the snakes, she abandoned her gear with a piercing scream and dashed up to the far end of the long corridor, where she cowered, with every sign of the most lively terror. It was no easy matter to calm her down enough to get her out of the danger zone, and hand her over to the caretaker for further treatment.

I thought this might have been an isolated case of rather specialized hysteria. However, a few years later, when I was a student at Marseilles, a similar thing happened to me in a pet shop owned by a lady. My errand was to buy a few small specimens of pythons that appeared in the price list of the establishment concerned. On the ground floor, next to the office, were various rooms which could be used as cages, since the doors had been replaced by wire netting. Several rooms were full of birds, tortoises, monkeys, *etc.* In order to inspect the snakes, I was shown into an empty room, in which was a sackful of newly-arrived pythons. An assistant willingly opened it, and shook the contents out on to the floor

Just as I was busy extracting a good-sized python from the writhing mass, I saw a female form walk past the wire-netting door, and at that very instant there came a scream, followed by a loud thud, and then groans. At once the whole house was in a regular uproar. The assistant threw down the snake he had just been praising and dashed out. Two negroes rushed past; one had a water jug and sprinkled the face of the manageress, who half lay, half sat in a corner, gurgling.

It appeared that I was the innocent cause of the incident, for how could I know that the lady, the proprietress of the shop, could not stand snakes, as they now reproachfully explained to me, with much gesticulating. Whenever she happened to see any, a similar outburst at once followed. Weaker reactions than this, though often accompanied by exaggerated expressions of horror and disgust, are often noticed in zoos when the visitors, particularly the womenfolk, see the snakes. I have also come across a case of a man, a politically influential newspaper owner, whose reactions on seeing snakes, even pictures of them, were extremely violent.

Perhaps the strange and little-known " imu " illness of the Ainu women of East Asia has some connection with the abnormal behaviour of these two women. Dr. K. R. Andrae has the following to say (1935, p. 697) : " The sight of snakes, especially vipers, which are much hated by the Ainus, playing a large part in their superstitions, frequently causes attacks of ' imu ' sickness. That is why it is often called ' tokkoni-bakko ' (tokkoni = viper)." According to Professor Uchimura's latest information, one or two women " imu " patients are to be found in nearly every community, even the smallest. This implies therefore a relatively common disease ; unfortunately, reliable records are not available. Between her attacks, the female " imu " patient gets on with her work unobtrusively, and leads a completely regular existence. For some reason, often inexplicable, she suddenly has an echolalic or echopraxic type of reaction. Either kind of reaction can usually be induced by saying, suddenly and suggestively, an irrelevant or unusual word during a trivial conversation, or by unexpectedly provoking the quite normally-behaved patient by word or gesture into performing a particular movement. Many patients react like a flash on seeing a snake, or the picture of one, or even if the word " snake " is said suddenly. With a look of extreme fear, they clutch at the nearest object to use as a weapon against everything around them, and nearly always show abnormal strength, considering their age and physical condition,

These attacks usually last a few minutes ; many patients remember nothing about them ; others recollect incidents and remain for a time deeply depressed. Cases have also been observed with a cataleptic reaction. In others, the attack is characterized by negativism. Whether, as the Japanese authors declare, this " imu " sickness is a case of a disease unique to the Ainu women, whose forms of expression are exclusively conditioned by the animal superstitions of the Ainu, remains open to doubt. More likely, this is a form of hysteria such as has sometimes been observed among other primitive peoples.

To return to animal reactions, let us again emphasize that here we have mentioned only a few special cases of what has sometimes been called by the vague and ambiguous term " animal hypnosis ". I agree with F. Steiniger (1936, p. 349) : " that at present it is impossible to give a full explanation of the essence and origin of ' animal hypnosis ', valid for all individual cases ; in fact the descriptions available today are only applicable to certain groups of single cases, and none of them, to the whole of the phenomena comprised by the term ' animal hypnosis '. There is in fact every likelihood that this summary of the phenomena includes quite heterogeneous material, that is in no closer reciprocal biological connection ; and that in future, no universally valid definitions are to be expected. Perhaps, in individual investigation as well as in the formulation of theories, it might be well to confine oneself to single groups of phenomena, to their occurrence within individual units of the animal system, rather than to plump as a matter of course for a universal explanation of all animal conditions comprised in the term ' animal hypnosis ', as was always attempted, especially by the first investigators of these phenomena."

And yet I still believe that the few indications given here, and the cases observed in person, can prove to us that, in the first place, there are all sorts of transitions between escape, with the swiftest possible change of location, and motionless shamming of death, as measures of defence against enemies. In the second place, it has been shown likewise that there are gradations between regular deception, *e.g.* the formation into lures of individual parts of the body on the one hand, and as organs to rivet attention (fascinatory organs) on the other. In the third place, the defence reaction of frogs or toads for example, which can be observed only under special conditions, so that effective forms of " counter-behaviour " exist for the most apparently refined fascination behaviour.

We have now reached a point to which I should like to draw the attention of the reader who is interested in biology, and on which I have already touched in the last chapter. We generally find a great deal of specialization in the predator, allowing it to succeed in catching its victim in a special manner. In the latter we find a like amount of specialization, effectively permitting it to avoid this same specialized chief enemy. The escape reaction, with appropriate equipment of the victim, corresponds to the prey-catching reaction and equipment of the predator, to some extent comparable to the positive and negative in the field of bodily structure and behaviour.

This would seem to present an extremely paradoxical state of affairs ; its interpretation lies in the province of natural philosophy rather than of animal psychology. Is nothing similar to be found in quite different spheres of nature ? In what infinitely varied ways the earth's surface is drained—and yet this same nature circulates this same water back on to the highest points of our planet. Strangely enough, this appears to us less surprising than that the victim's protective measures and escape reaction should be the counterpart of the teeth and claws of the predators, or the defence reactions of frogs and toads of the fascinatory manoeuvres of snakes.

In the zoo, where we are daily impressed, even affected, by the great variety of animal forms and ways of life, we are most forcibly reminded of the apparent paradox of adaptation and counter-adaptation. This is corroborated in nature in no uncertain manner. " Golden freedom " has two aspects ; one for the predator that is lucky enough to find a particularly tasty victim ; and another for the victim that is lucky enough to escape from a particularly dangerous enemy. One's zenith is another's nadir. Every wave crest has its trough ; to paint in rosy colours either the one or the other is to be guilty of anthropomorphic conduct. In the zoo, and perhaps outside as well, man's real task is to illuminate the shadows in the trough of the wave, and to shorten their duration.

6

ANIMALS AMONG THEMSELVES

As A RULE, we do not realize how much our daily life is bound up with strange customs and habits, until we are suddenly confronted with ways in other lands to which we have first to get accustomed. With us, in Switzerland, the gentleman walks on the lady's left, in France on her right ; in Scandinavia to drink on one's own from one's glass is considered impolite. The further away a country is, the funnier its customs seem to us. Anyone who is invited to dine with a high-born Arab in Morocco must remove his shoes before entering, and never on any account ask after the wife (or wives) of his host, who sits next to him at dinner, but refrains from eating a morsel himself. In addition, it is considered a breach of good manners to fail to belch audibly at the end of the meal.

Just as every nation has its own customs which often seem strange to us, so every species of animal has its own special social ceremonial. In contrast to man, the ceremonies of animals are largely innate, not inculcated. Man can learn afresh, the animal hardly ever. If we wish to behave properly and correctly in our dealings with animals, we must acquire a profound knowledge of the innumerable ceremonies of the animal world. Not only does each species of mammal, bird, reptile or fish have its own fixed and inborn ceremonial, but so do many invertebrates. One often encounters a surprising, even overwhelming amount of obligatory ceremonial among the so-called lower species.

Social habits of bees are proving to be of increasing complexity, as Karl von Frisch has recently shown (1948) in his excellent investigations ; the same thing applies to the mating ceremonial of the common fruit fly, *Drosophila*, so annoying to us when they settle on our fruit, and which are bred in their thousands in every biological laboratory (U. Weidmann, 1951).

The mutual relations, social forms, and recognition behaviour among animals of the same species comprise an enormous field of research which is the province of animal sociology, and has in fact only been seriously studied recently. The results of research in animal sociology are not of purely scientific and theoretical interest, but are often of prime importance for the management of zoological gardens. If we wish to handle an animal properly, be it a fish, a song-bird or a giraffe, we must know about the rules and ceremonies essential in each case for intercourse between animal and animal.

It is hardly necessary to stress that we are only just beginning to enter this field ; for there are at least a million species of animals, each of which may have many ceremonies. It is not surprising if we can discern certain relations between animals, but not yet explain or identify them. In other cases, rare it is true, we have progressed enough to be able to meet the animal correctly with its own specific ceremonial, and can thus treat it as would a fellow member of its own species ; the dog, for instance, whose social organization has much in common with man's. Possibly this has helped it to become man's first domestic animal.

I should first of all like to refer to some examples of relationships between animals that we can merely record, but not yet understand. For examples I shall draw on the tropics, partly because there the animals live much closer together than in our latitudes, and also because it is there that, as a roving scientist, one can find more time for observation than in our daily mechanized life, taken up with various duties at fixed times, when observations are likely to be constantly interrupted by so-called urgent business. Three incidents particularly impressed me in this connection in the South Sea Islands.

To begin with, there were shoals of finger-length fish, composed apparently of thousands of individuals. Such a shoal might be from twelve to fifteen feet in length. In the crystal clear water of the lagoon, every detail was visible. In the Bismark archipelago, I watched these fish time and time again as they manoeuvred, making right and left turns, but always with the hinder-most fish turning at the same moment as the foremost ones. Occasionally a thin, sharp-fanged spear-pike (*Sphyraenidae*) streaked up ; the whole shoal then leapt into the air three or four times, in such unison that it plopped back into the water as one fish. Never did a single one get out of order, or jump too soon or too late. Undoubtedly a strict and precise method of communication must have existed between these thousand fish. What it consisted of remains a complete mystery, even taking into account the fact that many fish can communicate with each other by colour or movement signals, by smell, or by the help of their lateral line system, a kind of organ of remote touch. Yet none of these explain their complete unanimity of reaction.

It was in the Solomon Islands that I saw the second example of synchronization, and an exciting one it was too ; that of an insect—a cicada. Just before the short twilight these two-inch long cicadas began their loud humming, with a crescendo like a dynamo reaching its maximum revolutions with a uniform hum. Many other cicadas do the same, but the ones I am speaking of made quite a loud crackling sound before starting the humming noise.

It was some time before I realized that it was the cicadas that regularly produced this crackling noise, before sunset. First I put it down to a frog, then to a bird. The solution, however, brought me no rest, for on my way back from a trip, or after heavy work in camp, just before the tropical twilight, which is always a fresh and thrilling spectacle, whenever I walked along a track through the virgin forest, I often searched for the cicada that produced the crackling sound.

Then a very strange thing would happen—the noise could not be located. Whenever I tried to get to the sound, creeping through the undergrowth to the place where I thought the insect was, I became undecided. For the noise, audible every ten or twenty seconds, seemed very distant, and indeed, to come from all directions, so that I could not trace the cicada, particularly as the crackling only lasted about a quarter of an hour each evening, to be followed by the universal humming.

At first I thought something must be wrong with my ears, nor would this have been surprising in view of the large doses of quinine we had to take every day to prevent malaria. Later it was discovered that the crackling sound did in fact come from all directions, and thus could not be located, since it was produced by a large number of cicadas at exactly the same moment, to a split second. I only solved the problem by accident, when I once had several cicadas in my field of vision at the same time. In some puzzling way, the cicadas must have been in communication with each other, to be able to begin their crackling sound at the same instant. Perhaps it was preceded by some preparatory noise beyond the range of audibility for the human ear.

But the most impressive, and at the same time the most delightful example of understanding between animals in my experience, and one which is still inexplicable, happened in a row of coconut palms on the little island of Umboi between New Britain and New Guinea. We were returning from an inland trip rather late at night, which was unusual for us, to the home of a planter on the coast, whose guests we were. The bungalow stood among tall coconut palms on the side of a hill rising abruptly from the sea. As we were about to climb the hill, following the curve of the coast, the whole row of palms was lit up by a bluish light for a few seconds. Then all was dark again for a short time until a fresh fairy-like glow bathed the landscape. This magic illumination switched on and off rhythmically. The light seemed to come from countless tiny points, and had gone on for some time before we recovered from our surprise and realized that it was caused by thousands of glow-worms flashing their light organs in mysterious unison, and magically floodlighting the palms on the hillside.

Any scientific, and that means critical, observer experiencing a phenomenon of this kind for the first time, especially in the often queer atmosphere of the tropics, must begin to wonder whether he is not the victim of a hallucination, or of an error of observation. Here there was no room for doubt, however. Yet I had misgivings about reporting this strange experience when I got back, especially as old and experienced tropical travellers whom I discussed it with out there, had never seen or heard of anything like it. So it was with relief that in the *Quarterly Review of Biology* (Vol. 13, 1938), published in Baltimore, I recently came across a complete study by John Bonner Buck, in which he makes a critical examination of all available reports on synchronized lighting, and illuminated insects.

It appears that as long ago as 1727, a detailed account of the phenomenon of instantaneous illumination by insects was given by a scientist, Engelbert Kampfer (1651–1716), on a visit to the River Menam area, near Bangkok, Siam. This particular observer rather naively assumed that his insects, that formed a huge cloud of fire, were rhythmically concealing and then exposing their tiny lights. Buck's critical account reports that synchronized illumination by insects has undoubtedly been confirmed in scores of cases, but that up to the present no satisfactory explanation of it is forthcoming. Nobody yet knows how individual glow-worms or, indeed, how most animals communicate.

Animal sociology is only just beginning to embark on its heavy programme of research, yet it has already shown that zoos, with their

Figure 8.—Giraffe begging in Rome Zoo (1952). For the animal psychologist, begging in Zoo animals is most interesting, as it involves types of behaviour not usually found in that form in freedom.

Figure 9.—*A moorland sheep, brought up on the bottle, which may almost certainly be said to have been ' imprinted ' (in K. Lorenz's sense) by its keeper, i.e., it always regards him as its mother and companion.*

Figure 10.—*Young female white-bearded gnu* (Connochaetes taurinus), *largely bottle-fed, and quite tame with its keeper, E. Glucki, even when fully grown.*

different groups of related and non-related animals, offer an extremely valuable field of work for the comparative sociologist. For instance, in 1947, Rudolf Schenkel produced a first class thesis on the communal social behaviour of wolves and their communication code. In it he shows, among other things, how very strict the wolves' social ceremonial is. The whole attitude of the body, especially the head, the position of the ears, the degree of baring of teeth, the bristling of certain portions of the fur and, in particular, the way the tail is held, are minutely prescribed. Apparently no movement of any kind is permitted unless it is in accordance with the social position of the animal concerned, otherwise there is a sharp rebuke. In fact, failure to obey the code may result in serious fighting.

The outsider can hardly imagine the complexity and ruthlessness of such social codes, not only among wolves, but in all animal communities, and even among species the individuals of which keep to themselves, when two or more individuals meet. Even if we entirely exclude intercourse between members of different species, there still remains a large number of categories of behaviour within the species which the animal sociologist should study in detail. C. R. Carpenter, the leading American field biologist and animal sociologist, has been working on extensive and painstaking research into the relations between animals ; *e.g.* the howler monkeys (1934), rhesus monkeys (1942), gibbons (1940) *etc.*

Let us take a simple case in Basle Zoo. At one time there lived on its monkey terrace some thirty-eight Java monkeys, two silver monkeys and a black male tufted mangabey from the Congo ; forty-one individuals all told. It is a well-known fact that each individual stands in a particular social relationship to every other in such a community. This relationship may be considered as reciprocal, *i.e.*, we assume that between monkeys A and B there are the same relationships as between B and A. Thus we may divide the total of all the individual relationships concerned by two. Let us reduce these facts to a formula, as C. R. Carpenter has done (1942a), where N is the number of monkeys. We can thus express the total inter-individual relationships as

$N\ \dfrac{N-1}{2}$, *i.e.* $41\ \dfrac{(40)}{2}$ = 820 different individual relationships on our little monkey terrace !

It would no doubt be most tempting to the animal psychologist to follow up this astonishingly large number of communal relationships. It is obvious, however, that this would mean a stupendous task. As an indispensable preliminary, each individual monkey would of course have to be made recognizable, and here, strangely enough, we encounter technical difficulties which have so far proved insurmountable. It would mean fitting every single monkey with a suitable collar or belt, with a number attached. But this apparently simple measure is attended with so many drawbacks that we have had to give it up.

Up till now we have contented ourselves with painting the leading monkey, the so-called α animal, with a recognition mark. This too has its drawbacks, as no paint can stand up for more than a few days to the monkeys' nimble fingers, and their intensive grooming. In addition, frequent catching of the animals for re-painting causes excitement and disturbance, and this disturbs the picture of the subtle inter-individual relationships under investigation.   Hitherto it has been customary in zoos to mark the queen bee in the insect-house apiaries with a spot of paint.   In this case, it is far more permanent than on a monkey's coat.

Unfortunately it is not possible to go into details here about the community structure of species of animals that have been investigated in nature or in zoos. We can merely state, for example, that there are patriarchally and matri-archally organized animal communities, *i.e.*, those in which the leader, or α animal, is always a male, and others in which it is always a female. Here we must be careful not to allow ourselves to be deceived by the imposing impression given to us by some animals.   Anyone untrained in animal psychology who had to decide which animal occupied the α position in a herd of ibex, would certainly declare straight off, on seeing the powerful horns of the male and the female's insignificant ones, that the much more impressive male was the α animal.   The truth, however, is that the he-ibex is the perfect hen-pecked husband, and it is the female that wears the trousers, so to speak.   Not only is this clear from field observation, but it can also be repeatedly confirmed in the zoo.   Thus, technically speaking, the female ibex is dominant over the male, since before he can get his heavy horns ceremonially at the ready, he will have had a couple of stomach prods from the nimble short-horned female.

Here we would like to make special reference to some of the regulations that control relations between animals and that are valid over a very wide range, from fish to mammals, even including many invertebrate groups. There are for instance animals that when at rest arrange things so as to get into closest possible contact with other members of their species.   Such animals are fond of being stroked, and are "contact animals".   They include wild boars, hedgehogs, porcupines, many monkeys and lemurs, tortoises, owls, *etc*.   This has nothing at all to do with the nature of their skin.   Animals with silky hair may be of the contact type just as much as the prickliest or the most bony-plated.

Similarly, it is not possible to recognize "distance" types externally ; it depends on a very special mental make-up.   Distance animals avoid close contact with animals of the same species.   Tame members of this group withdraw before the human hand, and dislike being stroked or scratched. They show the so-called "individual distance" very clearly* which varies according to the species.   In the black-headed gull it is some 12 in., in the flamingo at least twice that, in the swallow barely half the distance.   When dealing with animals, it is important to make allowances for their contact

* In connection with this term, P. J. Condor in *The Ibis*, the quarterly journal of the British Ornithologist Union, Vol. 91, No. 4, 1 October, 1949, pp. 649–655, has made a slight error.   He states that the term "individual distance" was first used by Dieter Burckhardt in 1944.   Without wishing to attach too much importance to this matter, it might perhaps be appropriate to point out that this concept, which was soon widely adopted, especially in ornithological circles, was coined by myself, and first published and defined in 1941, p. 43. By way of illustration, I used a photo of black-headed gulls (*Larus ridibundus*) taken by myself in Bellevue Square in Zurich in March, 1938.   Several gulls were sitting on the parapet by the lakeside, and as it happened, each was at a distance of two embankment railings from its neighbour.   On this occasion, I hit on the idea of individual distance, which is valid not only for birds, but for many other animals as well.

or distance character, as the approaching hand may be a welcome caress to the one, but make the other shudder.

Mistakes are often made in this respect, not only in the practical handling of animals, but in artistic or educational exhibits as well. A few years ago I was in the rich and lavishly equipped Natural History Museum in London, looking at that part of the bird collection where bird flight is strikingly displayed. A single mistake from the animal psychologist's viewpoint struck a discordant note in this exhibit, which is a combination of model, photographs and stuffed birds. In one scene in this impressive exhibit were two stuffed swallows, sitting on a wire so close to each other that the feathers of their adjacent wings touched. But this sort of thing only happens to stuffed swallows, never to living ones, which have a pronounced individual distance, once they have left the nest as fledglings. This mistake has since been put right.

In the zoo, with so many animals from every branch of the zoological system and from all parts of the globe, it is vital to group them as suitably and as significantly as possible. Respect for the laws governing intercourse between animals is essential.

In studying societies of mammals, it is of course necessary to realize the limitations of animal society, especially in two directions, external and internal. The classical subject of animal sociology, the typical animal society, is the group structure of animals of the same species.

Let us consider briefly the limitation of external social phenomena. It is clear that this cannot be ascribed to the barrier between species, since there are numerous social relationships in groups of animals, the elements of which are composed of different species ; this is even more true for artificial communities set up by man, to which he himself often belongs. Animal sociology should not ignore such phenomena between animals of different species, especially since hybridization is linked up with highly complex inter-specific social relationships. A border-line case is represented by that category of behaviour between animals that I described as " biological rank " in 1940. Here we are concerned not with rank for single individuals, but for whole species or races whose geographical ranges and biotopes overlap, in part at least, and which show a certain similarity in their physical organism, being thus on a footing of biological competition with each other, *e.g.*

> ibex > chamois > deer
> grizzly bear > black bear, or leopard > cheetah
> spotted hyena > striped hyena, or gorilla > chimpanzee

The species mentioned first is biologically dominant (the α species), the next named being subordinate (the β species), *etc.* The result of this for the single individual is that a representative of the subordinate species has to give way without fighting to a representative of the dominant one, if a question of space or food arises.

There are border-line cases in which the relationship of biological rank may develop into one of predator to prey, as for instance among seals. Normally, in the population of the Northern seals, the walrus is the α species, before which the different seals, as representatives of subordinate species, must give way. On the other hand, it sometimes happens that the walruses behave as predators, treating the seals as prey, and devouring them.

From the point of view of animal sociology, external limits are fixed by biological rank. The predator-prey relationship lies beyond it, and belongs in fact to a very different sphere of functions, that of the enemy, not that of social relations.

On the other hand, there are further inter-specific relationships bordering on the sphere of problems of animal sociology ; we need only mention certain kinds of symbiosis, *e.g.* among different species of ungulates.

We will quote only one example of symbiosis which I managed to observe, apparently for the first time (Hediger, 1951), at the Garamba National Park in the Belgian Congo a few years ago. It concerned the hippopotamus (*H. amphibius*) and a certain fish some 18 in. long (*Labeo velifer*). As we were watching a herd of hippos at the confluence of the Garamba and Ako rivers in April, 1948, I was struck by the fact that large black fish kept swimming sluggishly round the hippos, and nibbling them all over, face, eyes, corner of the mouth, *etc.* When one of the huge hippos surfaced, a couple of these fish might be seen clinging like leeches for a few moments to the hippo's body, then dropping off. The mouths of these fish, called by the natives dorumbia, were leech-like. It is possible that the fish, which we later saw again in the Dungu River, clean the skin of the great mammals, perhaps eating the sticky sweat, algae or sediment. To our surprise, this apparent symbiosis between animal and fish was well-known to the natives, who called the dorumbia fish " water cattle heron".

After this digression we must return to our theme, the discussion of the subject of animal sociology. Phenomena of animal sociology are thus limited externally, as we have shown by biological rank on the one hand and by varieties of symbiosis on the other.

Let us go into the possibilities of limitation internally. Here too we can start from the normal basis of animal sociology, *i.e.* from the structural group of individuals of the same species. One might think that at least two individuals are necessary for manifestation of social phenomena, yet these may even occur in single individuals. The so-called " lone-wolf "—usually an old male—is a familiar and striking example among many mammals. Exceptional phenomena of psychological deficiency may arise through artificial solitary confinement, and these are without doubt of very considerable sociological interest. The " odd-man-out " type is not confined to human beings. It is a commonplace among zoo biologists that single animals often lose their appetite, and that food intake is materially increased by the so-called social factor (Hediger, 1950, p. 122), *i.e.* by association with a companion.

I should like to mention one border-line case, in connection with the sociology of mammals, and that is the extreme case in which an individual comes to terms with parts of its own self. Up to the present, I have been investigating two cases of this sort, which are naturally very rare among mammals. Among lower invertebrate organisms, on the other hand, they are extremely common, *e.g. Hectocotylus*, starfish and in particular sponges. There are even sponges that can be strained through a fine sieve, when the separate portions, often only single cells, re-unite into one organic sponge body ; this brings us up against a sharp dilemma over the concept of individuality.

The disintegration of the individual mammal indicated does not of course go as far as this. Nor am I thinking of tissue cultures or other artificial conditions, but of completely natural phenomena : the process of birth, on the one hand ; the shedding of antlers in stags on the other ; that is, of treatment of the afterbirth and of the discarded antlers. At first, this may sound rather surprising. Yet the behaviour of mother and child is clearly of sociological concern, and therefore the mother's treatment of the embryonic sac and the afterbirth are very closely connected with it.

As a result of his fine work on free-living monkeys, C. R. Carpenter (1942, p. 180) recommended the study of eleven categories of inter-animal relationships. This classification is also very useful for other mammals, so I should like to use it, slightly modified, as a guiding principle for these explanatory remarks.

Generally speaking, these are the categories concerned :

1. Male : male
2. Female : female
3. Male : female
4. Male : young
5. Female : young
6. Young : young
7. Group : group
8. Solitaries : group
9. Inter-specific relations
10. Relations to space
11. Individual : group

In this outline, I should like to add a few notes as extra sub-divisions on the social significance of antlers (12), the assimilation tendency (13) and begging in animals (14).

In contrast to the human psychologist, the animal psychologist, as we have said, is faced with the extremely difficult situation of having to deal not with a single species, but with an almost unlimited number of different species. The mammalologist alone has at least 3,500 species to master ; the behaviour expert, who deals with this highly developed class of animals, would have to know the same number of behaviour patterns as well. Thus the device to classify into types, into main and subsidiary groups of particular behaviour categories, is easy to understand. In 1941, I made an attempt to blaze this trail and formulate a few universal types. Some of these have since been adopted, more have been added. Moreover it appeared that certain types of behaviour might often be very simply expressed in symbols borrowed from geometry.

1. Male : male

Fighting is a dominant behaviour-type in full-grown males of the same species. It is the last resort in settling disputes as to which is superior, which inferior, after the various methods of display or threatening have failed. Now among mammals, these threatening gestures are as a rule better known than the actual fight behaviour. That is why I should like to consider a few examples from the group of ruminant ungulates, and take the opportunity of showing that definite connections exist between different types of horns and antlers and the corresponding actions.

Not all horned ungulates use their horns for attacking their opponents. Most caprinae and ovinae possess typical striking weapons. Both opponents stand facing each other and take a considerable run in order to strike each other with the full force of their horns, either directly, in a horizontal direction or else vertically, striking in a downward direction.

According to V. Schmidt (1935, p. 26), male mouflon sometimes take a run of twenty yards or more. " The noise of the impact is audible a long way off and sounds like lumberjacks at work. After the clash, both rams step quickly back and the run begins again. If the opponent loses his balance or is knocked out, the victor strikes the cervical vertebrae of his victim with his spiral horns in order to break its neck." Thus, as in the case of the stag, the battle takes place along a straight line, each fighter attempting to throw his opponent off course and so to expose its unprotected flank.

Other sheep fight like the mouflon, for instance the African maned sheep recently studied by I. Katz (1949) in the Bronx Zoo, or the thick-horned sheep that I was able to observe in San Diego Zoo. In contrast to the mouflon, *Ovis canadiensis* goes in to attack erect on its hind legs for the whole of its run. During mock battles between rams of this kind—and between ibex as well—one of the contestants may chance to stumble in its run up, or may not be quite ready for some other reason ; accidents like these cause the partner to pull up short and start again.

In Basle Zoo we were frequently able to observe a type of behaviour, which we have called pseudo-contact, between rival rams of the domestic sheep (*e.g.* Lüneburg Heath sheep). When the flock was let out of its narrow pen into the meadow, at first the two almost equally powerful rams used to fight every day. After a time, the rams began to push each other sideways so hard on their way out that neither of them was able to take the decisive run. Which of the two took the initiative in this pseudo-contact it was impossible to discover. The rival pair would run round the field like a pair of weird Siamese twins until something happened to break this strange antagonistic bond.

In contrast to the sheep family, the horns of the goat family are not held quite parallel to the ground during fights, but are aimed more or less vertically from above before the impact. In September, 1952, we lost a male goat of the Wallis canton breed at Basle Zoo, through a broken neck. In a fight against a ram of the Lüneburg Heath sheep, the goat, while it was parrying the horizontally aimed thrusts, forced itself into a position unfavourable to its anatomical structure. This example illustrates the close connection between behaviour and physical build. In the chamois (*Rupicapra*) the horns are never used as striking weapons, but exclusively for ripping open. Horn functions are very varied in antelopes, and so is their fighting ceremonial. Many antelopes, instead of striking from above, thrust up from below, often dropping like a flash on to their metacarpal joints. These tactics are very common among short-horned species, *e.g.* the nilgai (*Boselaphus tragocamelus*). The male okapi, too, thrusts upwards with its short horns. Antelopes with long sabre-like horns fight in the horizontal, not the vertical plane, and sometimes kneel as well (*e.g.* Oryx ; Hippotragus). It has long been known that some Bovidae, the gaur and the gayal for example, attack not frontally, but laterally, *i.e.* broadside on, thus using only one horn.

These few examples are quoted simply to show that in fights between rivals there may be a connection between shape of horns and behaviour. This connection was closely studied by H. Bruhin (1952). For the sociologist the problem not yet solved is why in some species of ungulates with horns or antlers both sexes have frontal weapons, while in others only the males. In this connection the Belgian mammalologist, Serge Frechkop (1946, p. 4), drew up the following rules : " Among horned ruminants, both sexes bear horns or antlers in the case of species, the representatives of which live in large herds, and in which the females do not outnumber the males, or only slightly ; as for example Rangifer, Damaliscus, Syncerus, *etc.* On the other hand, only the males are horned in those species whose representatives live in small groups consisting of one male and several females." Here too there appears to be an intimate connection between morphological and sociological appearances.

Elsewhere I have tried to outline a rather different interpretation of the variety of horns or antlers in male and female ungulates (1951). I started from the fact that horns that are not periodically renewed are by no means of equal importance. These include the mechanically effective weapons previously mentioned, for example the powerful horns of the Bovidae or wild cattle (American and European bison, buffalo, *etc.*) as well as the sword-like horns of the oryx, antelopes and gnus, the strong, medium-sized horns of many goats or the dangerous ripping hooks of the chamois.

Horns of these kinds, obviously of effective mechanical use against enemies of a different species, thus providing efficient means of survival value, are found in both sexes. Males and females alike may have the opportunity of defending themselves against predators.

Apart from these, there are horns so slender that, on purely technical grounds, their use as weapons against enemies of other species is obviously out of the question. They may only be used as light weapons when the opponent has no more efficient or more solid equipment, but about the equivalent, and uses them according to fixed rules, *i.e.*, with specific ceremonial. This is exactly the case in fights between males of the same species. Horns of this sort, characterized by their delicacy, even fragility, are usually found only in the males (*e.g.* Impallah, Adenota kob, Pantholops).

Naturally, highly developed specific fighting ceremonies also occur among mammals without horns. It is of interest to note in this connection that teeth, used as weapons, reach their maximum development among pure herbivores and not among predators, at least as far as land animals are concerned. The record goes to the elephant and the hippopotamus. The huge constantly growing canine teeth of the hippopotamus are typical social equipment, used for fighting between rivals. A fight of this sort is usually decided after a fixed ceremonial. Each opponent stands more or less parallel to its rival, but facing opposite ways. Then each one tries to drive its eye tooth into its opponent's flank by a powerful side stroke of the head. Cases have been known of the huge canine tooth being thrust into the heart of the defeated rival, and fragments of eye teeth are often to be found on the hippos' fighting ground. Recently H. N. Southern (1948, p. 181) described the fighting ceremonial of wild rabbits (*Oryctolagus cuniculus*), which jump at each other with such force that a distinct thud can be heard as they meet, often a yard or so above the ground.

Moreover, in fights between rivals of the same species, the same weapons as in fights against members of different species are not necessarily used. In fact, several weapons and special expressions may be reserved for each of these two functions. Thus for instance the giraffe does not use its horns in fighting off predators, but only its feet. The horns are reserved for social encounters. In fights with members of their own species, red deer lay their ears right back, but with members of other species, they point them forwards. There are thus not only distinctive social weapons, but also social fighting expressions, often quite independent of the weapons and expression that function in the sphere of the enemy. In fights between rivals, the rattlesnake (*Crotalus ruber*) uses neither its poisonous fangs nor its rattle : yet both are brought into action in fights against members of different species.

2. FEMALE : FEMALE

It is a familiar fact that among various mammals, males and females have special social rank. In the case of the American bison, a young male must first work its way up through all the female ranks, and then start again at the bottom of the male hierarchy in order to win itself a position. During the rut, a cow bison may stand practically at the social level of the leading bull, who for the time being never leaves her. I observed a similar state of affairs among brown bears.

Generally speaking, female animals are considered to be less pugnacious than the males. In organized animal fights, it is always the latter that take part—*e.g.* bull fights, elephant fights, cock fights or exhibitions of fighting fish. In Switzerland, even today, there is an uncommonly popular animal fight, that is, the cow-fight (" combats de vache "), in the Val d'Anniviers in the Wallis canton.

A small primitive race of cattle, the Ering cattle, have managed to survive there. Every year, in the second fortnight of June, at the time of the move up to Alpine pastures, these cow fights take place, and from time immemorial they have provided the local population with the chief event of the year. According to W. Gyr's account (1946), the fights between cows happen more or less like this. On the day of the move, the cows are driven from the various stalls up the mountains, where for the first time since autumn they meet again at traditional fighting places, *e.g.* on a meadow or in a forest clearing. These fighting places are dotted with trough-like depressions about one or two yards across. They originated through the cows in war-like mood pawing up the ground in preceding years. Every year on the way up to the Alpine pastures, the same fighting spirit returns and individual fights take place between two cows.

As a rule, each cow fights every other ; after several hours, often only after several days, a social rank has emerged. Inferior rivals suddenly face right about, make off, and give way in future to the superior animal. The α cow, the so-called " queen " (" reina "), usually keeps her position all summer. It has often happened that a queen has maintained her high social rank for several years on end. The steers that go up to the fighting places play hardly any part, since they are invariably young animals, subordinate to the cows. The queen is usually adorned with a particularly fine cow bell. It is not a fact that she always leads the herd when it moves on but she invariably uses her full authority, however, generally separating fighting members, as other α animals do.

3. MALE : FEMALE

The different forms of mating are among the earliest problems investigated by animal sociologists. Yet at the present moment there is very little material available on the preliminary courtship and ceremonial of mammals, particularly when compared with the abundant and detailed observations on fishes, amphibia and reptiles, and above all on birds (Armstrong, 1947). O. Antonius (1937) published a very important comparative study of the Equidae, and so far there is nothing comparable to it on any other family of mammals. In it he classifies the six groups of single-hoofed mammals : true horses ; *Equus hemionus* ; donkeys ; steppe, mountain and Grévy's zebras. From the sociological aspect, true horses have a definite herd formation, led and watched over by the strongest stallion. The typical facial expression of a donkey in heat is entirely lacking in the female horse. In *Equus hemionus* and donkeys, the herd is led by a female. The loose-knit herds of steppe zebras have no real leader ; mountain and Grévy's zebras form only small bands in which leadership is doubtful.

We must expect many surprises among the multiple groupings of even-toed hoofed animals in general, and of antelopes in particular. Above all there are the transitions from brutal chasing of females by males (*e.g.* chamois) to almost mute and motionless courting (*e.g.* ibex) and complicated ceremonial by pairs (roedeer) or collectively (impallah).

The ibex in rut does not chase his mate, but climbs his look-out, where he stands for hours with the typical rutting expression, *i.e.* head outstretched and nostrils wide open. From time to time he protrudes a quivering tongue, and emits a faint bleat. His tail is tilted up over his back. If a female in heat appears, the male swings round like a compass needle to face her, and may even make begging motions with its forefeet. He then sometimes follows her at a fixed distance until mating finally takes place.

In districts where the mating ceremonial of roedeer can take place undisturbed, the couple move in a circle during that phase when the male with his laboured panting seems to be driving the female to exhaustion. The female can in fact break off the apparently wild chasing and start it again at will. During this phase, the male is not driving the female in order to mount her ; mating later takes place without chasing. The circular tracks they trample out are sometimes called " fairy rings " in England.

Here we have described the ceremonial circling round of a single pair, but M. Verhulst, the former warden of the Parc National de la Kagera near Gabiro in the territory of Ruanda-Urundi has observed impallah antelopes circling collectively, no doubt in connection with rutting. Some seventy females and young were grazing together ; at one side stood a considerable herd of about twenty males. Now and then a couple of males fought face to face. Then, as if at a signal, all the males, heads outstretched and tails up, began running one after the other in a circle round the herd of females, emitting a typical roaring sound. This strange roundabout continued for about five minutes, and was repeated several times. The animals taking part gave the impression of being highly excited. At the end of this ceremony, which was later seen by other observers as well, the animals went on quietly grazing.

Apparently many corresponding ceremonies may be described for antelopes. D. Burckhardt watched the behaviour of the nyala antelopes in the Bronx Zoo, New York, which reminds one of the lateral displaying among pheasants. In this case, one can tell from the formation of the tail whether the type of pheasant concerned has a frontal or a lateral display. In the group of *Tragelaphinae* among the antelopes, to which the nyala belongs, the most striking thing, apart from sexual dimorphism of colour and horns, is that this antelope has various patterns on its flanks (stripes, spots, *etc.*) and that it can change its lateral contours to an astonishing extent by stiffening its mane, or the tufts of hair on neck and breast, and often too the specially thick hair on the tail or belly. Many of these *Tragelaphinae, e.g.* the Indian nilgau, have the characteristic habit of walking in slow motion round their females, and sometimes round their rivals too, and always showing their flank to them.

But in the zoo, no less than in freedom, it must be stressed that there are many kinds of relationships between animals which at present we can neither predict nor understand. Once, to our surprise, we watched something of this sort with members of the sheep-goat family. Thanks to building operations, a herd of Himalayan tahr had to be accommodated next to two male African maned sheep. One morning, a worried keeper announced that somehow or other a female tahr had got among the Barbary rams, but was as yet unharmed. The necessary steps were at once taken to get the Asiatic ungulate back among its own fellows. The tahr was particularly valuable to us since it was a mother suckling a fine kid.

Next day the same thing happened, although all communicating doors had been carefully secured. Meanwhile, before the keeper on duty could summon help for the straying mother tahr, she was walking around her own enclosure again as though nothing had occurred. She must have got back on her own, but how she did it remained for long a mystery. These visits to her African neighbours became eventually so frequent that the game would be up sooner or later. In fact, thanks to a stroke of luck I managed to photograph the female tahr as, without taking a run, and without visible effort, she jumped from the top of the rock elegantly over the six foot gate, only to return to the herd in order to suckle her kid, which could not of course follow its mother over such obstacles.

Far more astonishing than this skill at high-jumping, which is common to a surprising degree to many mountain ungulates, is the fact of animal sociology, that neither of the Barbary sheep did any harm to the Himalayan intruder.

It is quite common among mammals, during the rutting season, to see an extraordinary physical proximity between partners. Ungulates, and others, may follow each other during this period without intermittence often literally step by step. Later, this close approach gives way to normal distance once more (social distance : see below). It seems appropriate to speak of the typical almost constant proximity during the rutting season as the specific rutting distance. C. R. Carpenter (1942, p. 189) regards this short distance in rhesus monkeys as an infallible sign of the onset of heat.

4. MALE : YOUNG

The relations between mammal fathers and their young are of very contrasting kinds. Every gradation from active care of the young (wolves) to the brutal treatment reserved for prey (polar bear) are found. According to S. W. Harms (1948, p. 226), it even happens that in some bats, the males carry the young alternately with the females. In addition to the account of circling ceremonial previously mentioned, I should like to say something about the very different " fairy rings " found by the first settlers on the North American prairies. According to Garretson (1938, p. 40), they originate when the bull bisons continually circle the cows during the period of the calves' birth, to protect the latter from attack by coyotes. When the calf is a few days old, it is able to rejoin the herd with its mother, where it is protected. Here is a case of active defence by the father, typical among other cattle, such as caribou, banteng, gayal and yak, as Antonius says (1933, p. 154). In contrast to this, the father in many other ungulates, e.g. red deer, chamois, ibex, etc. is never even present at the birth.

On May 19th, 1950, in Basle Zoo, a zebra mare, that had been put by itself in a stable as a precaution, gave birth to a delightful foal, and this caused the stallion in the next box some excitement. For reasons of space, but also because it seemed biologically sound, we wanted mother and son to meet the father next day in the inner stall. At first all went well, and we were just congratulating ourselves that the whole family would settle down nicely together. Then the unexpected happened. The group had scarcely left the stable and gone out into the paddock, when the stallion fell like a lion upon his new-born son, knocked him brutally over, and mauled his neck, like an animal with its prey.

We had been prepared for all kinds of incidents, but not for such a brutal attack. With whacks from brooms and kicks, we eventually managed to get the raving father off the ill-treated foal before he killed it. Against wild animals that have iron-hard hooves, a vice-like bite, and the temper of an enraged tiger, brute force has sometimes to be used if one is to make oneself understood by them. Naturally, the stallion was immediately separated. That evening, when we put the mother into the box along with the foal that had been rescued in the nick of time, the zebra stallion, freshly excited, did its best to jump over the six foot six partition, and banged around its stall like one possessed. By rearing up it managed to get the hooves of its forefeet over the top several times. Zebras might well be called tiger-horses, not because of their striped coats, but because of their predator-like rages.

For the sake of peace and quiet, we had to transfer the zebra stallion with all precautions to another house, in this case to the antelope house, where at first it ran against the iron bars, but after the cage had been screened, it calmed down again in a few weeks. Six months later, on November 22nd, 1950, when the foal had now grown up satisfactorily, we repeated the attempt, under the necessary safeguards. This time, all went well ; father and son greeted each other cordially, so to speak, and since then the family has been united heart and soul, if one may be permitted the expression.

If I were asked to explain this infanticidal behaviour of the zebra father, I should confess to finding it quite impossible to give the right answer. I can only offer a conjecture. The whole affair impressed me as if the zebra father, in his excitement at anticipating his protective functions as head of a family, had mistaken his offspring for a threatening enemy of another species, and had acted accordingly. Such things do happen. But this pre-supposes that the stallion in this species, namely Grant's zebra, actually plays the part of an α-animal, of a leader and protector, as is certainly the case with the wild horse. All we know at present among the many species and varieties of zebra—and there are dozens of them—is that, sociologically speaking, some behave like horses, others like donkeys, and others again are half way between the two. Thus it will be necessary to know more about the sociology of the zebra before forming an opinion on the present poor attempts at an explanation. But apart from O. Antonius's fragmentary accounts (1937 and 1951) already mentioned, there has not so far been any serious research.

About two years later, on June 22nd, 1952, the same zebra mare gave birth to a foal of the same father, also a male, weighing 83 lb. Some months before its birth, we had isolated the pregnant mare at night, and for months on end the night watchman paid a visit every few hours in great anxiety to the zebra's stall. He was not lucky enough to witness the birth. Mares are notorious for extreme secretiveness in bringing their foals into the world. Recently, in the *Zeitschrift für Tierpsychologie* (1951), W. Koch has shown that mares have the unusual faculty of delaying the hour of birth for some time, at least twelve hours.

Obviously, zebra mares can do the same thing. At all events, the fact that the said mother Grant's zebra gave birth to her foal just between the night watchman's last round and the keeper's first, *i.e.* between 4 a.m. and 7 a.m. is evidence in favour of this.

Mindful of the unfortunate experiences we had with the zebra father in 1950, this time we almost overdid things in our attempts to exclude all risks. We decided to bring about the mother's reunion with the herd and the foal's introduction to it in easy stages. For several days, and for a short time only, we released mother and son in the adjacent paddock, so as to accustom the herd gradually to the novel sight, and with a stout fence between them. On the first occasion, the whole herd, led by the father, crowded to the fence in the greatest curiosity. The mother, cautiously and on the defensive, placed herself between the fence and her foal. She showed much skill, too, in manoeuvring him with head and body movements, in the desired direction, *i.e.* away from the stallion and the fence.

After several repetitions of this the next few mornings, and with a ring of keepers armed with brooms on the qui vive at all strategic points of the enclosure, the herd gradually became accustomed to the sight of mare and foal in the next paddock. Eventually we felt able to take the risk of letting the mare into the herd again by herself for a short while, once more under all possible precautions. A certain amount of rough galloping and chasing around took place, but without any serious biting, and no neck-breaking, and things gradually settled down to normal.

Not until July 10th, two and a half weeks after the birth that is, did we venture to introduce both mother and foal to the herd, and then only for quarter of an hour. It was amazing to see how clever the mother was at shielding her frightened foal, kicking out in his defence with her hind legs. In time, they got used to it and settled down. The young male foal was quietly accepted into the herd and gradually precautions could be relaxed until finally, after about three weeks, every risk seemed to have vanished, as far as it was humanly possible to judge.

With most bears, which are considered not without reason as symbols of matriarchy, the father is in no circumstances allowed into contact with the new-born cubs. The female polar bear retires to a hole in the snow and ice with a very narrow entrance, and defends it against the father with might and main, since the cubs' lives are at stake. In zoos, once at Zurich for instance, polar bear fathers have been known to break forcibly in and devour their own young.

In contrast to the Ursidae, the males in the Canidae often take an active part in rearing their young. Within one family of animals, however, the relations between father and offspring may be on opposite lines, e.g. in the Leporidae. In the case of wild rabbits (Oryctolagus cuniculus) the doe digs a special burrow, the so-called nursery, a mere blind tunnel, in the warmly-lined wide end compartment of which the young are born. The burrow is carefully sealed by day and only visited at night by the mother. In the common hare (Lepus europaeus) on the other hand, the father is present at the birth and apparently assumes the duty of active protection. In the course of scores of births in captivity, we have never taken the males away, and in no single case have the young been injured. Similar behaviour occurs in the wild pig (Sus scrofa). The boar has often been said to devour its young whenever it could get hold of them. On the other hand, after many births in the zoo, I have come to regard the father wild boar as very peaceable and remarkably well tempered. Once on March 29th, 1951, I was able to photograph one that had even installed six of its young on its own back. I discovered that the male African porcupine (Hystrix cristata) is a completely fearless defender of its young ; it would sometimes take its baby between its forelegs, or even tuck it under its belly to protect it, rattling its spines and bristling. The mother showed far less interest in it.

The role of the father among anthropoid apes has not yet been properly determined. Apparently he undertakes some protective functions, and this is the case in certain of the lower apes. The extent of paternal help clearly depends on personal experience as well. At the birth of their first-born, they often behave clumsily and nervously. Thus in 1941, A. Urbain, the Director of the Paris Zoo, and his assistants observed at the birth of an orang-outan that the father was obviously embarrassed when the mother invitingly laid their offspring in his arms. It was some time before the male orang learnt how to take the baby from its mother from time to time, and hold it properly.

When, moreover, this baby orang died aged eight and a half months, the keepers had the greatest difficulty in removing the corpse from the mother. This is a common experience with monkeys in zoos, and can also be confirmed among free-living monkeys. Dead babies are carried around as if they were alive by the mothers for days on end, even when they show obvious signs of decomposition. One of the zoo's most repulsive tasks is the hunt after such dead baby monkeys.

The family types (Parents—Mother—Father—Family) outlined by H. M. Peters (1948) for fishes have their counterparts, no doubt, among mammals.

How extraordinarily varied the organization of the family, and especially the role of the father, can be is very vividly illustrated in the zoo by peacocks and emus. The peacock territory—a model for demonstration purposes, with its internal arrangements—shows in addition to all the types of fixed points such as sleeping tree, playground, sandbath, feeding place, resting area, *etc.*, typical display grounds as well, used by the peacocks year after year, apparently for many generations.

In accordance with his space and time system, the peacock displays his tail-spread at these places at fixed times (March to July) and thus advertises that peahens ready for mating can be served there. This takes place with highly characteristic ceremonial, fixed to the very last detail. As Heinroth puts it (1938, p. 64) the peacock " holds his treading hours ". It might be mentioned, by the way, that the peacock cannot be provoked into displaying his tail either by whistling, stone throwing, waving a red rag or suddenly opening an umbrella, *etc.*, in spite of the fixed beliefs of most zoo visitors. It is rather a question of biological action arising from their specific space and time system.

The only part the peacock plays in bringing up the family is when treading the hen. He takes as little trouble about nest-building as he does about brooding or management of the young, all this being left to the hen exclusively. She broods the eggs for about a month without once being relieved. During the course of the morning, she leaves the eggs for about twenty minutes to attend to the needs of nature. Once, quite exceptionally, I saw a young peacock squat down right alongside a brooding hen.

Almost the exact opposite is the case with the emu. The hen bird just lays the green-shelled 21 ounce eggs, not always in the nest scrape, and leaves the cock to hatch them out. Unlike the peahen, however, the male emu does not even give himself a short daily break, but sits about two months on the eggs, fasting and thirsty, and growing correspondingly emaciated. He merely changes his position once or twice a day, rearranging the eggs. Later the cock also bears the brunt of rearing the striped chicks. Still, he is helped by the hen in protecting them against marauders, amongst whom, sometimes, they include the zoo keepers.

There is every conceivable gradation between the complete failure of the peacocks to share in rearing the brood and the emu's one hundred per cent participation. The mute swan, for example, does not sit, but often stands protectively over the clutch, while the Australian black swan takes regular turns at brooding. From morning until late afternoon, the cob is on duty ; then the pen takes over. With African ostriches, the shifts are reversed, the grey hen brooding from about 9 a.m. till 6 p.m., when her black mate takes his turn at brooding until the next morning. These spells are very unequal ; many small birds change over every quarter or half hour ; some vultures, however, change only every couple of days or so. In my opinion it would be a most fascinating task for an ornithologist to tabularize the methods of brooding and the turns of duty relief for as many species of birds as possible. No doubt plenty of interesting relationships would emerge from such a piece of work for the student of behaviour.

5. FEMALE : YOUNG

Naturally, among mammals the relations between mother and young must be far richer and varied than in any other class of vertebrates, since it is only in their case that this decisive dependence on the mother's milk occurs. It is thus advisable to devote a special chapter to the varied forms of this relationship.

6. YOUNG : YOUNG

In connection with the relations of the young to each other, I should like to touch on one phenomenon of the biology of ungulates, to which closer attention should perhaps be paid. We see from the case of the wild pig, that among ungulates that produce a large litter, all the young are born at the same place, in the same hole. Many ruminants have only one at a birth ; among those that regularly have twins, however, it may often be observed that the two youngsters are not born at the same spot but at some distance apart. In such cases one might use the term twin-distance, and this is no doubt as specific as other biological distances.

In roe deer, this twin distance consists, to my knowledge, of no more than twenty yards as a rule, while in the American pronghorn (*Antilocapra americana*), according to E. Heller's observations (1930, p. 4), it is about a hundred yards and according to A. S. Einarsen (1948, p. 107) even more than 130 yards. After being suckled, the young are sometimes left a similar distance apart, up to a certain age. In the case of roe deer, this sometimes has the familiar but disastrous result that people come across the fawns while on a walk, and thinking they have been abandoned, take them home.

The zoo visitor enjoys the delightful sight of groups of young of many ungulates, especially deer, whenever there are a number of them. The fawns that are born about the same time, and which keep apart at first, join up at a few weeks old and practise their running and jumping games at definite times, usually towards the evening. Fallow deer fawns especially, their tails sticking stiffly up, are in the habit of dashing in close formation up and down the whole length of their paddock, and this they repeat time and time again. Such a company of fallow deer is very striking. It might in the terminology of P. Deegeners (1918, p. 277) be described as a " sysympedium ", although strictly this term applies rather to a combined group of offspring of several mothers.

7. GROUP : GROUP

Here too I should like to confine myself to mentioning one characteristic distance, the group or herd distance. Usually among socially organized groups of mammals there is a marked tendency to keep a definite distance from one another. In many cases this distance is already determined by the fact that each group lives in its particular territory, which is demarcated specifically. In those species, however, in which no such strong spatial ties exist, that is to say in those in which several groups move about more or less freely in a considerable area, the herd distance is very clearly marked. H. Krieg (1940, p. 96) has described this herd-distance, *e.g.* for flocks of guanaco, consisting of six to twenty animals, the so-called " cuadrillas ", which live under the leadership of a male on the vast grass-lands.

I should now like to refer briefly to an observation on the Cape hunting dog (*Lycaon pictus*), since the relations between individual groups have not been at all clear to me. Thus in 1948 we encountered a group of seven hunting dogs, joined half an hour later by a further group of six animals, after a peculiar wolf-like ceremonial. The leading dog was easily recognizable by his behaviour and was also physically distinguishable by a torn left ear. Three days later we came across a pack of sixteen Cape hunting dogs at the same spot, under the same leader. Obviously a certain amount of breaking up and amalgamation of groups occurs with these grim and seldom seen predators. It would be worth while making a closer analysis of these relations.

8. SOLITARIES: GROUP

I have already mentioned the solitary animal, the one that has become secondarily a-social, so often encountered among mammals, and which cuts itself off actively (by seclusion) or passively (by expulsion) from normal social ties with its species. As a rule there should be no such thing as solitariness in a zoo, since keeping individuals singly is fundamentally unbiological (Hediger, 1951). On the other hand, one is sometimes in the position of having to look after single individuals exceptionally in practice, for instance, when one of a pair has died, and cannot be replaced or because a certain animal is a danger to others and has to be isolated. In the zoo it is therefore necessary to distinguish between intentional isolation, through external circumstances, and that caused by an animal's behaviour.

The only kind that interests us here is that caused by the animal itself, that is to say, biologically motivated isolation. It is not without parallel in freedom where it occurs, as we have said, among the most widely different species, that certain individuals, usually males, are expelled from the normal family or social unit on account of their behaviour. F. F. Darling (1952) supposes that in the case of the solitary red deer (*Cervus elaphus*), the cause is usually atrophy of the gonads. I do not feel, however, that it is safe to generalize from this fact.

Of the larger African wild animals, elephants and buffaloes are among the most notorious—that is dangerous—solitary animals. Both have this in common, that old elephants and old buffaloes usually have prominent weapons in the shape of tusks or horns which cause them to be picked out specially for shooting by ambitious game hunters. This fact, and the loss of collective protection formerly offered by the herd, lead, in such solitary animals, to a sort of chronic state of super-irritation and alertness. To this must be added the negative experiences they have had at man's hands.

Now comes a most astonishing fact, observed in the case of the buffalo, which is rightly considered as the most dangerous wild animal in the African continent. When it gets into a reserve, as for example the Albert National Park in the Belgian Congo, this same solitary old buffalo becomes extraordinarily familiar towards man. In consequence, as I myself have proved with a tape measure, its flight distance is sometimes reduced to the seemingly ridiculous figure of thirteen yards.

There, for example, in the " Camp de la Rwindi ", either on foot or in a car, one can encounter old buffalo bulls whose behaviour is not essentially different from that of cattle in the Swiss Alps, so familiar are they. This is probably due to the following factors. These animals have been expelled from their natural herd unit, and are consequently in a continual state of defensive irritation against their fellows. They have to be much more on the alert against their natural enemies too (*e.g.* lion), than the rest of their species that have remained together and warn each other. They have in fact lost that protection offered by society.

When they reach a reserve, they quickly learn that they are no longer molested by man, and also that the closer they keep to him, the freer they are from aggression by their own species. At any rate, this seems at present the best explanation of the fact that from being by far the most dangerous of solitary animals, in the reserve, they develop into the most faithful " hotel pets ". The same thing may be seen in the zoo where individuals that are socially ostracized and chased away, attach themselves to man with special intimacy when isolated. Only recently we saw the same thing again in a socially inferior female baboon, which was never able to bring up her baby with the pack, but only when she was eventually isolated.

9. INTER-SPECIFIC RELATIONS

Under the title Mosaic Patterns (1950, p. 17) I described the vast field which the study of the literally countless inter-specific relations offers. One must take into account the fact that within its habitable area, each species of animal, so to speak, fills up one territory after another, so that a proper mosaic is formed (cf. Chapter 2). But this mosaic of one species is overlaid and permeated by many mosaics of other species. In view of the large numbers of possible relationships, it is advisable to describe not only isolated examples, but categories of behaviour types. In my opinion, J. M. Linsdale (1946) in his monograph on the Californian ground squirrel (*Citellus beecheyi*) has made a most attractive analysis of this mosaic pattern in connection with a particular species of mammal. Linsdale went into the relations between this rodent and many other vertebrates, including thirteen species of mammals, which may be divided into the following three main categories :—

1. Predator-prey relationship (*e.g.* with *Mustela, Taxidea, Canis, Felis, Lynx,* etc.)
2. Biological rank (*e.g. Citellus > Eutamias*)
3. Mutual tolerance (*e.g.* to *Sylvilagus*).

It is clear that with different species, many other types of relationship occur, especially positive ones of a symbiotic nature, as have already in part been described by Alverdes (1925) and Deegener (1918).

I should like to refer briefly to one peculiar case—a border line—since to my way of thinking it illustrates a fundamentally important factor. In the summer of 1949, I made the experiment in Basle Zoo of keeping some cattle heron (*Bubulcus ibis*) not—as usual—in the aviary, but at liberty, in the natural company of larger mammals. Normally, these African birds spend a large part of their day perching on all kinds of large animals, *e.g.* elephants, buffaloes, hippopotami, and rhinoceroses, as well as on domestic animals, especially cattle and sheep.

I put five of these birds in a field occupied by Lüneburg Heath sheep, Wallis canton goats and South American capybaras (*Hydrochoerus capybara*), giant rodents weighing up to one hundredweight apiece. I had a definite feeling that the experiment, which was made, incidentally, with cattle herons that were almost completely capable of flight, would succeed. The first bird that I released in the field flew up into a tall tree about 150 yards off. The second I plunged into the water, and then let it swim across a small pond, so that it could not fly straight up. On reaching the bank at the other side, the cattle heron spotted a capybara, watching it. The bird went straight over to this animal, and it was plain to see that it felt quite at ease in this giant rodent's company. For their part, the capybaras allowed the hitherto unfamiliar bird to approach. Since some sort of short-cut contact had clearly arisen between these dissimilar creatures, I released the remaining three cattle heron in the same way. They all went straight across to the capybaras and even the first that had flown off came down from its lofty perch and joined the rest.

From that time on, the cattle heron never left their capybaras the whole summer, although they could fly and sometimes took short flights round about. Very often they would perch unconcernedly on the resting capybaras, allowing themselves to be carried about pick-a-back. When one of the capybaras that was lying on the ground, rolled over luxuriously, the bird riding on it took the necessary steps to keep on top.

The astonishing thing about this artificial community of animals that had never seen each other before, some from South America, others from Africa, was first its instantaneous origin, as soon as the animals had sighted each other, and the mysterious understanding with which either partner met the advances of the other. On the part of the cattle heron, there was naturally a certain need for contact, but the fact that their behaviour corresponded at once so completely to an obviously latent response from the capybaras was most striking to observe.

10. RELATIONS TO SPACE

Following on the discovery of the surprisingly close connection between most birds and definite sections of space or territory, it has been found that this principle is also valid for most mammals. The territory of mammals is variously demarcated (cf. Chapter 2), and the marking of it is almost exclusively a male prerogative. This finds its anatomical expression in the fact that in the males the scent glands are far more highly developed.

Hitherto I have known only a few examples of bi-sexual territorial marking. One is the short-tailed East African mungo, and another the serval. Demarcation by both sexes seems to me to occur especially among animals which show no striking sex dimorphism.

In some wild dogs, representing rare cases of non-confinement to a territory, true nomads that is, the rudiments of space fixation have been transferred to the females. This is so in the Cape hunting dogs, for example, and the South Asian wild dog (*Cuon javanicus*). Cape hunting dogs appear quite unpredictably in the landscape, then vanish again, so that one wonders where and when they rear their young. In contrast to the wolf, the males have no obligations whatever during the birth of the young. The pregnant females assemble at some favourite place in a hollow and rear their young collectively, only to resume their nomadic existence as soon as possible afterwards.

11. INDIVIDUAL : GROUP

According to its species, each individual keeps at a greater or lesser distance from its group; that is, the group shows specific social distance. Carpenter (1942, p. 184) for instance reports that the pack of howler monkeys is more compact than that of spider monkeys. Thus, as I see it, the latter has a greater social distance than the former. This concept is of importance since it may have practical diagnostic importance in the running of a zoo. Separation of an individual from the pack is often the first symptom of illness. The same thing may be seen in birds. In flamingos, excessive social distance is a dangerous symptom. In female mammals, an increase in the social distance may be the first sign of an impending birth, as we have often very clearly seen with zebras. In exceptionally threatening situations, a concentration of the social unit may occur, namely a sharp reduction in the social distance, as has for instance been observed by A. S. Einarsen (1948, pp. 166 and 197) during the opening of the shooting season for the American pronghorn (*Antilocapra americana*).

Social distance also plays a part in dog management. Dogs with too great a social distance are just as troublesome as those in which it is too small. The former cannot function as companions, the latter are no use at scouring a piece of country.

Rutting distance, already mentioned, presents to some extent a special case of social distance ; the latter is always greater than the rutting or individual distances. Social distance means the maximum distance between individuals of any group; individual distance on the other hand denotes the minimum distance within which individuals may approach each other.

12. SOCIAL SIGNIFICANCE OF ANTLERS IN DEER

In contrast to most other mechanical weapons (teeth, hooves, horns, claws, *etc.*) of mammals, the antlers of deer are distinguished by periodical shedding and by their high social symbolism. It is known that as a rule only stags with antlers of approximately equal strength will fight. Those with weak antlers never face battle. Yet it does not follow that one can tell the social standing of a stag just by the size of antlers.

It is common knowledge in zoos that stags which have shed their antlers have at the same time dropped in the social hierarchy. With red deer, and some other species in which the older ones generally shed their antlers before the younger ones, even the highest in rank are attacked as soon as they have lost their antlers. Here it must be remembered that some time before shedding, when osteolysis is already considerably advanced, the antlers can play no further part as mechanical weapons. Yet the attacks by socially inferior individuals never occur until after shedding. Thus during this time, at least, the stag is protected by the symbolic value of its antlers.

It has been objected that the attacks to which the stag is subject after shedding its antlers are perhaps not due to their loss as a social symbol, but have been provoked simply by the weakened condition of the animal when under the shock of shedding, in the same way as any physical weakness may provoke members of its own species to attack. I could show that this line of argument is untenable from the case of three male Sika deer (1946, p. 162).

In this trio, a stag of top social rank (α) was vigorously and repeatedly persecuted by one of very low status (γ) as soon as it shed its antlers, and it had to be transferred to an adjacent enclosure, in which was the β-stag, which had shed its own antlers twenty-three days previously, and was then frequently attacked by the α-stag. On being transferred the newly arrived α-stag immediately set upon the β-stag, although its antler scars were still bleeding. There can be no question of shock after shedding antlers here, since a stag with shock if possible retires and never spontaneously attacks, least of all on strange ground. Besides, Sika deer have comparatively small antlers, in contrast to reindeer or red deer, in which shock effects might sooner be expected.

The attacks of the γ-stag on the α-stag just after the latter had shed its antlers can therefore only be explained in this case from the fact that the α-stag had lost his social status with his antlers. Thus antlers are a highly significant symbol for the animal, so long as they are carried in the typical manner.

The observations of F. F. Darling (1937, p. 160) and others have shown that certain relationships may also arise between antlers and their former wearer once they have been shed. According to them, these cast-off antlers are at once eaten by the stag in districts that lack calcium phosphate. This is one of the rare cases of the re-use of part of the body that has mortified, and it is to some extent on the border-line of sociology.

Let us add that the stag's antlers, like the elephant's tusks, may also be used as an erogenous structure for a peculiar kind of auto-eroticism, since they are concerned with a partial vacuum activity of the mating ceremonial.

13. ASSIMILATION TENDENCY

The history of behaviour study is at bottom the history of the struggle against the deeply seated tendency in every human being to humanize the animal. Even the biologist must carefully guard against the intrusion of anthropomorphisms; the more primitive the man, the greater his humanizing tendency. He humanizes—*i.e.* demonizes—not only animals, but plants and even completely inanimate objects as well. We can see how alive in us this rooted tendency to humanize is even today by glancing at the advertisement posters or newspaper advertisements, where for example we can see sausages or machines with human faces.

This proclivity, in the form of an animalizing tendency, is even more strongly marked in animals. Animals' zoomorphism corresponds to man's anthropomorphism, both being manifestations of the assimilation tendency. Here we are concerned only with the social consequences of this characteristic, so deeply rooted in every human being and animal. It also has the result that a mammal with which we are on close terms, *i.e.* which is tame, and for which we have no enemy significance, often sees us as fellow members of its own species, and consequently accepts us too into its social unit. The man who has to work with animals in a practical sense must thus play the role of a fellow member of the species; he must win for himself a social position, must assert it, and above all must be conscious that he may become the object of rivalry or of attempts at mating. All this is true even for the elephant. It is clear that anyone concerned with the handling of animals in zoos and circuses must reckon on this fact in dealing with his charges.

It is sometimes rather hard to differentiate between this assimilation tendency and Lorenz's " imprinting " (" Prägung "), although he originally thought that the phenomenon of imprinting should be ascribed only to birds (1935, p. 166) and insisted on the criterion of never being able to forget. True, assimilation often occurs very slowly in mammals, especially with older animals. Sometimes, in young mammals, however, it may arise suddenly as a " fixation of the urge to follow ", *i.e.* a new-born ungulate may follow close on the heels of man, his horse or car, just as it would follow its mother.

As early as 1867, accounts and illustrations existed (cf. Garretson, 1938, p. 40) showing how very young calves of the *Bison americanus*, when cut off from their mothers, followed the huntsmen's horses so closely that they sometimes got kicked. At the age of one or two months, this urge to follow disappeared, however. The African game-hunter Berger (1910, p. 105) had a similar experience with a *Syncerus* calf, that even went hunting with him. It proved quite impossible to get rid of this animal with its man-fixation. L. Heck (1930, p. 98) when in Africa happened to ride past a new-born zebra. " The baby foal jumped up, ran behind our car, and was literally not to be got rid of. The mother had dashed off, the youngster had failed to contact her, and trotted around after us."

Among different varieties of domestic sheep, we often had the following experience with lambs that had been brought up on the bottle, because of the failure of their mother's milk. For some time after weaning they would leave the flock and run up to the keeper whenever he appeared. F. Goethe (1939, p. 6), likewise mentions " imprinting " in his report on the artificial rearing of two young mouflons (*Ovis musimon*) which had an " irresistible impulse " to follow the girl who looked after them.

In this connection, W. H. Hodge's account (1946, p. 656) of a cross between an alpaca (*Lama pacos*) and a vicuña (*Lama vicugna*) is most interesting. This cross-breeding has never been observed in a zoo. It can only occur when a new-born male vicuña has been caught. The Peruvians kill at the same time a new-born alpaca and cover the young vicuña with its skin. The latter is then accepted by the mother alpaca and reared. Only vicuñas of this kind later pair with alpacas and produce paca-vicuñas, *i.e.* hybrids. In other words, the vicuñas must be " imprinted " on alpacas, and then become quite assimilated and pair.

In the spring of 1952 I made a curious observation on the new-born female twins of a Lüneburg Heath sheep in Basle Zoo. The lambs, born in March 1952, were unequal in strength and weight; 4¾ lb. against 4 lb. 2 oz. It soon appeared that only the stronger one reached the udder. The weaker one was repeatedly put to the ewe by us, and also fed from the bottle. Our hopes that this weak lamb would be accepted by the mother were disappointed. In fact, on the third day, it was obvious that the mother was forcibly rejecting the lamb. Finally the keeper isolated it, and brought it up on the bottle, with the result that it was soon dogging his footsteps all round the zoo, through all the animals' houses and the service quarters. It showed no interest in its mother, nor in its fellow animals, nor in any strange human beings either. After a year or more, it was still devoted to its foster father, so that even K. Lorenz, on his visit of October 23rd, 1952, was inclined to consider that it might be a case of " imprinting ", in spite of the fact that the lamb was not isolated from its mother and sister until after its third day. However, a vital loosening of the bond—a complete break, in fact—occurred, when the milk bottle ceased to be offered to the now grown-up sheep.

These few instances may be enough to show that even in new-born ungulates there is a tendency to fix their impulse to follow on a representative of another species in a way reminiscent to some extent of the " imprinting " that is typical for certain birds. There are probably different types of social attachment in mammals and in birds, and in other animals as well, among which learning, on one hand, and " imprinting " in Lorenz's sense on the other present extreme cases. One would not be very surprised to find that definite parallels existed among the autophagi, whether mammals, birds or insects, in connection with the mother-child relationship.

14. BEGGING

Apart from the assimilation tendency, which sometimes appears in a grotesque form, intimate contact between man and animals, so typical in zoos, produces another category of social relations. This is not found in free life, at any rate in the same form, and is called " begging ". Since this kind of social phenomenon is almost purely anthropogenous, it seems extremely attractive to me from the animal psychology point of view.

The fact that begging is not so widespread in the animal kingdom shows that here there is no question of primitive behaviour. Typical beggars are only found among mammals, that is to say among the highest ranks of the animal world. Innate begging reactions in birds, especially young ones, are on a very different plane. Hungry crowding and jostling when waiting for food, often seen among fish for instance, cannot be called begging in the true sense.

Genuine begging happens chiefly in situations in which a decline in possessions or position or both takes place, when it may assume the character of a handicap. At any rate, it cannot be overcome directly, so that begging is the expression of an attempt to conquer this obstacle, which incidentally may be physical or mental, real or imaginary. In different species, the begging organs vary a great deal. Naturally the extremities play a big part in it, while mimicry, gesticulating and the general bearing of the body are brought into play. From the tip of the nose (in elephants the end of the trunk) to the tip of the tongue (*e.g.* giraffe) or of the tail (spider monkey), the most varied organs may be called upon.

As far as the motive for begging is concerned, it would be wrong to think that only food is begged for. Other motives may be the struggle for a home, contact, companionship or sexual activity.

The goal at which begging motions are aimed, *i.e.* the object which is begged from, is by no means always a human being. Animals also beg from one another, as did the ibex previously mentioned. They sometimes even beg from inanimate objects. A polar bear that we were trying to entice into an inner compartment with a piece of meat stood stubbornly beneath the open trap door and begged at the attractive morsel. A pony with decayed incisor teeth made begging motions at the bucket, the contents of which were only used for cases of toothache. H. W. Nissen and M. P. Crawford showed in 1936 that chimpanzees beg coins off each other which they can then use to get food out of slot machines.

MOTHER AND CHILD

AMONG the most delightful sights for the zoo visitor is watching young animals, and the mothers' care for them. One always notices favouritism for young animals on the part of the public. Everywhere, they want to feed only the babies, and to entice the adults away from the food by all sorts of manoeuvres, even though the young are still being suckled, and have no need for solid food, while their parents could do with plenty of it.

During building operations, it once happened that our draught horse, which is not normally on view, had for the time being to be put up next to a pony's enclosure. Both begged in the usual way, but the horse nearly always went without, for most visitors lavished their tit-bits only on the nice-looking little pony, although it was very old.

We should now like to mention a few striking phenomena from the comparative point of view as they occur in the zoo, in order to demonstrate from them the possibility of classification into categories or types. Once again we shall mostly use simple geometrical terms. It is well known that nearly all the primates carry their babies during the early days, clasped to the abdomen. Here two fundamental positions may be distinguished: according to my experience hitherto, the monkeys carry their young parallel to the body axis. Many prosimians however carry them transversally. Lemurs, lorises, *etc.*, carry their young like a belt, slung round the abdomen. The only exception seems to be the ring-tailed lemur (*Lemur catta*), described by W. C. Osman Hill (1952), and which we were able to observe very well in the monkey house at Zurich Zoo. The picture published by the above-mentioned writer also shows, however, that this lemur sometimes adopts the transverse position. From a purely " technical " standpoint, this is not easy to explain, since both types have very similar organic structures. A parallel is found among bats, to mention only one more instance of it, as F. P. Möhres (1951, p. 659) has pointed out. Whilst the young of all native species hang head downwards, in the same direction as their mothers' body, young horseshoe bats (*Rhinolophus*) hang head upwards, clutching their mothers' genital region ; the so-called " clasping teats ", which must probably be considered as modified mammary glands.

I have noticed in young hoofed animals that, when sucking, their bodies have to form a definite angle—usually acute—with their mothers' body. This angle position is usually taken up in a particular manner, often by the calf circling the mother from behind, and then taking the teat at the correct angle from the front. Sucking positions are not random, but according to rule. This is sometimes overlooked by artists, usually such keen-eyed observers. For instance, on the Swedish 5 oere piece, a sucking foal is shown at quite an impossible angle.

In young ruminants whose mothers, with few exceptions, suckle them standing, it is obviously innate for them to find the teats they seek in the angle made by the vertical and horizontal parts of the mother's body. This innate pattern is somewhat irregular, for it sometimes concerns the front extremities and the chest, as well as the hind extremities and the abdomen.

Whenever I was able to watch new-born ruminants, or wild boars, trying for the first time to find the source of milk, I was always struck by the fact that at the start they looked for it in the wrong place, namely in the angle formed with one of the front extremities. They only learnt to differentiate between the right and the wrong place by experience, though this did not take long.

The suckling position of the mother, too, is specific. As we have said, nearly all herbivores suckle while standing. The only two exceptions I have seen so far are two telemetacarpal deer—the elk (moose) and the roe deer. With these, I was frequently able to confirm that in the first few days the calf or calf twins are suckled in the lying position, the mother raising her uppermost hind leg so that it does not get in the way. Of course, abnormal suckling positions can be caused experimentally. A white-bearded gnu (*Connochaetes taurinus*) born in Basle Zoo on July 31, 1950, sucked its mother for two months while she lay on the ground, as she had gone lame at its birth. The baby gnu was bottle-fed in addition.

The lying suckling position is specific for pig species and, significantly, for their nearest relatives, the hippopotamus and *Choeropsis*. These types of behaviour seem extraordinarily persistent, from the philogenetical point of view. There are various positions among rodents. The capybara (*Hydrochoerus capybara*), by far the largest rodent, suckles its young standing. In this species too, remarkably similar to ungulates, the new-born young sometimes mistake the proper place, *i.e.* they occasionally go to the throat area for the teats. The hare suckles her young sitting up and the sucking young sit facing her in a similar position. In the case of rabbits, however, the mother lies on her side, more or less on top of the young, which generally lie on their backs.

Just as the mother's suckling position is specific, so too as a rule is her position during delivery. Elephants and many ungulates (*e.g.* giraffe) drop their young standing up ; others (deer, cattle) usually lying down. I have watched exceptional cases of standing births in deer, for instance, and bison. With the hippopotamus, moreover, delivery usually takes place in the water. The ninety-pound baby must first kick its way up to the surface before it can draw its first breath. When it wants to suck, it dives down to its mother again, as she lies at the bottom of the shallow water where they live. Later she is used as a raft by her growing baby, allowing it to lie on her partly submerged back.

The search for the milk supply starts with ungulates as soon as they are able to walk, generally from quarter to half an hour after their birth. In the case of the long-legged giraffe, for obvious reasons, it lasts about twice as long—up to an hour—when it finally succeeds in hoisting itself up on to its stilts. In plural births, the first-born is sometimes sucking before the rest of its brothers and sisters have come into the world, as for example, wild boar. Suckling is not in this case interrupted by labour. Recently, from the gynaecological point of view, special interest has centred upon research on comparative lactation in wild animals, with regard to the most suitable biological suckling method for human beings (W. R. Merz, 1948).

Whilst new-born monkeys are put to the breast by their mothers, babies of other species (*e.g.* ungulates) must make their own efforts to get the milk, not only by active search, but also, for instance, by prodding the udder with the snout (Bovidae) or by massage, so frequently met with among rodents and predatory animals, and consisting of vigorously kneading the milk glands with the help of the forepaws. One may at times see experienced ungulate mothers helping their young to find the source of the milk supply by pushing them towards the teats with their heads. This I have definitely confirmed for Grant's zebra, white-bearded gnu and nilgai. Mother seals often use their front flippers to help their babies to find the right place.

Sometimes the mother has to use supplementary aids, if irregularities in normal behaviour occur. In new-born anthropoid apes which are unable to breathe, the mother may sometimes blow into the baby's mouth, and keep on until it starts breathing. R. M. Yerkes (1934, p. 266) reports cases of this nature. Once a mother chimpanzee, whose baby was not breathing, pulled its tongue out with her lips and blew until the little one started to breathe.

Dr. E. Bronzini of Rome Zoo told me he had often watched Californian sea-lion mothers grab their young by the nape of the neck when they failed to start breathing, and dip them repeatedly quite deep in the water, then lay them on land again, where breathing at once started. In Basle Zoo we thought at first that the mother should be kept on dry land during delivery, and lost several baby sea-lions soon after birth. A post-mortem in these cases showed inflation of the lungs to be non-existent or partial. From this it seems fair to assume that the breathing of the newly-born sea-lion is stimulated through this maternal treatment by immersion. At any rate, we promptly saw to it that our pregnant sea-lions had plenty of water available. Indeed, the baby born in 1954 was successfully reared.

I have repeatedly seen for myself how ungulate mothers tapped their weakly young with their forefeet, if they had not got up by the normal time, *i.e.* within half an hour. This encouragement clearly brought about fresh efforts on the part of the baby. In one case I watched a chamois mother carefully put her horns under her weak kid, and try to raise it that way.

Since many kinds of possible physiological disturbances arise in the development of the embryo, in the same way the danger of qualitative or quantitative departures from the normal may occur in every single element of behaviour in the mother–child relationship after birth. These are not always due to under-functioning, as for example insufficient lactation or defective rearing of the young, where the close connection with hormone metabolism has in many cases been proved. In the zoo, over-exaggerated care of the young, a partial hypertrophy of the maternal instinct, may sometimes be as fatal for the offspring. Predatory animals, for example, that try to hide their young in more and more suitable places of safety, dragging them around by the scruff of the neck for this purpose until they eventually succumb to exhaustion or even to serious injuries, and are sometimes even devoured, unless the mother can be given an adequate feeling of safety and security in time, by the removal of all disturbing influences. I remember very clearly examples of this kind among spotted hyenas and racoon dogs. In both cases, successful rearing of the young followed when an adequate amount of security could be provided.

Among zoo inmates, a considerable difference, as far as the mother–child relationship is concerned, may often be seen between the behaviour of experienced mothers and those with their first-born young. This distinction is often indeed so sharp that the first birth might be considered as something like a dress-rehearsal, not counting for the propagation of the species, but a preparation for subsequent births. Efficient rearing of young often seems inadequately developed in primipares ; they are not mature enough and the young are not fully developed either. There is a lack of that intimate dovetailing of the behaviour of mother and baby so that the weak baby soon dies, or is even still-born.

On May 26th, 1952, in Basle Zoo, we had the opportunity of recording a typical example of this kind, the first time a baby giraffe was born. The calf was underweight (68 instead of about 110 lb.), and far too small (4 ft. instead of from 5 ft. 4 in. to 5ft. 7 in.), and the mother was clearly afraid of it. As there are all too few details of giraffe births, I quote from the records :

Notes on the Birth of a Giraffe

9.00 a.m.   Keeper Glücki reports that " Susy " the giraffe shows signs of restlessness.

12.00 a.m.   Glücki thinks birth likely during the dinner hour interval.

12.30 p.m.   Inspection by myself; nothing to worry about before 2 o'clock; keep within reach of a telephone call.

2.00 p.m.   Glücki notices nothing on returning to work.

2.15 p.m.   First micturation on the verandah. Glücki puts Susy into the inner stable and rings up.

2.30 p.m.   On my arrival, about eleven inches of the forelegs visible. Distinct whitish eponychia. A lot of sand and straw hastily put into the large stall, which I think more suitable on account of the sweeping neck movements.

2.48 p.m.   With the tip of the calf's snout visible, Susy lies down.

2.51 p.m.   Stands up again. The birth process would obviously be prevented by lying on the ground.

2.53 p.m.   Cud-chewing—defaecation.

2.54 p.m.   Head protrudes about eight inches. Calf blinks; breathing movements perceptible.

2.55 p.m.   Head completely free; movements of calf and mother. Young giraffe tumbles out ; falls to the ground. Mother stands stock still.

3.2 p.m.   Mother takes first look at calf. Birth complete reverse of normal, which would have produced a right-hand turning movement.

3.3 — 3.5 p.m.   The mother, her mane bristling, circles the calf from left to right, and touches it several times with her front legs. First-born!

3.6 p.m.   Calf tries to raise its head for the first time. It is still partly wrapped in the amnion.

3.8 p.m.   Raises its head; its neck collapses. Manages to hold its head up for a second.

3.11 p.m.   First effort to stand up.

3.12 p.m.   Sudden gush of amniotic fluid, second portion. Mother circles round calf from left to right.

3.13 p.m.   Calf turns round; raises its head for second time.

3.15 p.m.   Tries to stand up; topples over forwards.

3.18 p.m.   Mother feeds from rack; ignores calf completely.

3.34 p.m.   Turns to calf for a moment; continues to feed.

3.35 p.m.   Mother walks round calf again, left to right; looks at it from time to time.

3.40 p.m.   Another attempt to stand, but tumbles down head foremost.

3.54 p.m.   Repeated attempts to get up.

4.5 p.m.   Calf still quite untouched by mother.

4.45 p.m.   Mother attempts to catch the attached placenta by sweeping movements of her head.

5.00 p.m.   For some seconds, mother looks more closely at the calf.

5.7 p.m.   Mother drinks with legs astride from drinking trough.

5.20 p.m.   First real contact between mother and calf; muzzle to muzzle.

5.28 p.m.   Second prolonged contact with head of calf, particularly its nostrils.

5.30 p.m.   Legs astride, the mother at last starts to lick the calf a little.   Keener interest gradually awakens: she circles her calf dangerously close. Risk of trampling on it!

6.5 p.m.   Still unable to stand unaided.   Glücki sets the calf on its legs for the first time and it is then noticed that the calf's right hind leg is broken, fractured by the mother treading on it with her foreleg, so that it cracked audibly.

9.30 p.m.   Calf removed from mother, as she obviously has not taken to it.   Calf drinks one-third pint fresh colostrous cow's milk.

10.5 p.m.   A further one-third pint of fresh cow's milk.

Subsequently, the baby giraffe was brought up on the bottle (fresh cow's milk), the mother showing no interest in it whatever. There is no doubt that during its first night, this underweight first-born calf would have fallen a victim to a predatory animal.

I have been able to confirm a similar state of affairs for many primiparous wild animals; immaturity on the part of the mother, and its typical fear of its offspring.

A female chimpanzee in the care of E. Steinbacher (1941, p. 192) in Frankfurt Zoo, failed through nervousness to cut her first-born's umbilical cord. First of all she allowed the baby to drag the after-birth around with it. later she carried it around in her hand for hours until eventually it had to be forcibly cut away. In the unique anthropoid ape farm at Orange Park, Florida, where the third generation of chimpanzees has been reared, R. M. Yerkes (1948, p. 68) observed that chimpanzee mothers seemed quite scared of their first babies, not even daring to touch them.

H. Heck (1940, p. 16) thinks that still-born first baby elephants are a more or less normal phenomenon. " It would seem that the first-born is often meant only to prepare the mother's body for the task of reproduction." On May 8th, 1932, the first Indian elephant was born at Munich—Hellabrun. At 2 a.m. the keeper on night duty suddenly heard a deafening row coming from the elephant house. He ran in and switched on the lights. " In the corner of the front compartment lay a baby elephant, apparently discarded by the angry mother. She had broken her chain and was lashing at her baby with it till the sparks flew, and trumpeting." The injured baby had to be taken away from its mother and artificially reared. They could not risk letting the mother suckle her baby until twelve days later. Moreover, it was thought that the mother was only eight years old when she gave birth.

In Rome Zoo, too, a twelve-year-old elephant, at her first birth, would have nothing to do with her 225 lb. calf that was born on August 6th, 1948, and it had to be artificially reared. This was the first successful attempt at rearing an elephant completely artificially. The composition of the milk given with the bottle—2 gallons a day plus ¼ gallon of tomato juice—was the result of an analysis made by Professor Anselmi of the Ministry of Health.

With elephants, birth takes place, as we have said, in the standing position; it is a remarkable feat that the calf is often presented hind legs first. According to A. J. Ferrier (1947, p. 77), it has been confirmed in Burma that the mother does not take any notice of the calf until the after-birth has been expelled. Normally the calf weighs about 225 lb., and is nearly three feet in length. In the case of an unusually tiny calf, the mother will often stand over a hump of earth, so that the baby can reach the two teats situated on her chest. Elephant mothers like to keep their calves between their legs for protection, where they can also support them.

By way of contrast to the "immature" behaviour of mother animals towards their first-born just mentioned, I should like to quote the records of the birth of a white-bearded gnu by an experienced mother. Two details should be specially noted—the fact that within 25 minutes the gnu calf was not only able to walk, but to hop and jump as well, and also that the mother pushed it with her head to the source of milk.

Record of the Birth of a White-bearded Gnu
June 16th, 1952

6.55 a.m. Telephone call.

7.15 a.m. At the Antelope House. Head and foreparts just emerging from the standing gnu mother, who now lies down.

7.17 a.m. Expulsion going smoothly, with mother lying on her right side. She gets up immediately; licks the calf's hindquarters; its head is still enveloped in the amnion. No breathing movements—threatens to suffocate. Sudden convulsive movements by the calf result in rupture of the head covering; very active; bleats.

7.20 a.m. Calf raises head for first time. Eyes open. Few patches of amnion on right shoulder. Rest of amnion already swallowed by mother.

7.23 a.m. First attempt to stand up.

7.24 a.m. Second attempt; hind legs are splayed backwards.

7.25 a.m. They lick each other's snouts.

7.26 a.m. Third attempt to stand up. Topples over. Eponychia shrivel.

7.27 a.m. Momentarily stands, its four legs wide apart.

7.28 a.m. Stands upright for the first time. Extra sand and fresh straw laid.

7.29 a.m. Short walk.

7.30 a.m. Stands until 7.33. Licks door posts. Looks for teats under mane on mother's neck; then at her rear; then again in wrong place. Feverishly searching. Repeatedly pushed towards right place by mother's head.

7.34 a.m. Stands up again, falls down, gets up at once and again looks for milk.

7.39 a.m. Looks in right place, then in the anal-vaginal region. Walks steadily after mother but has not yet suckled.

7.42 a.m. Again sniffs door post. Now frisky.

7.43 a.m. Mother lies down on her left. Calf frisks around her. 30 seconds later, mother gets up.

7.45 a.m. Still trying the wrong place; ears erect now.

7.47 a.m. Tries at right place, but too high up; then once more under the tail, and under the hanging placenta.

7.49 a.m. Calf leaps up, pointing its tail at the same time; tries the anal-vaginal region again.

7.50 a.m. Sucks for the first time, from between the hind legs, under the tail, and the placenta. The mother quietly allows it to, then walks away.

7.52 a.m. Again tries at the wrong place, sidles back against its mother, and then for the first time energetically sucks in the normal way. Mother sniffs the calf's anus. Calf lies down for a while.

1.00 p.m. According to keeper Glücki's report, placenta ejected, and in part swallowed.

While watching births in ungulates, it has also struck me, in connection with the first reactions between the mother and the young animal, that there are clearly two types of behaviour to be distinguished, which might be described as the active and passive types.

The active type, as far as my experience at present goes, is found in the single-hoofed ungulates, and most ruminants. It is characterized by the fact that the mother, who normally gives birth lying on the ground, stands up suddenly after the final expulsion of the offspring, turns right round and, with the avidity of an animal of prey, falls on the amniotic sac and devours it. In this situation, the most decided herbivore turns carnivore all of a sudden. I saw bison and antelope cows swallow such large mouthfuls that I was afraid they would choke.

This sharply defined reaction in wild animals is no doubt very practical ; it prevents the baby animal from suffocating in the amnion. In domestic animals that for generations have relied on man's help at birth, this useful reaction has largely disappeared, not only in ruminants, but in the seemingly much less domesticated donkey as well. In one case, a donkey mare gave birth to a completely normal foal, whose amnion was not ruptured during birth. The mare stood by the foal wrapped up in the amnion, without doing anything about it, and let the foal suffocate. Several similar cases of ponies are known to me.

This, in a wild animal, would be quite unthinkable. Here selection works pitilessly ; individuals that fail to rupture the amnion of their young, provided these do not break by chance, remain incapable of propagating their species, and the best guarantee of opening consists of the quickest possible and most complete devouring of the sac. The amniotic fluid is also often licked up by wild animal mothers to the last drop; if necessary, the umbilical cord is bitten through, and the placenta swallowed also as a rule. It is likely that the animal thus assimilates important elements.

At the same time, or immediately afterwards, the mother licks the baby animal dry. That is obviously not just an expression of deep " maternal love ". This licking represents a vital massage. The stimulation of the anal region by the tongue is of special importance, as many young animals are unable to empty either bowels or bladder without this stimulative friction. In artificial rearing, colic with fatal results often ensues unless this massage is replaced by rubbing with a fine sponge or rags. During the birth of a male deer calf, I was able to see the penis not only massaged but taken into the mother's mouth, so that the new-born baby's bladder was sucked empty.

In non-ruminant ungulates, especially pigs, but also in Camelidae (e.g. llama), this sort of reaction—immediately standing up, turning round in a circle, and swallowing the amnion—does not seem to be innate, but here, quite different behaviour patterns are found. Pigs (Suidae) usually produce a greater number of young at birth ; the mother neither frees them from the amnion, nor bites through the umbilical cord, nor licks the young dry. In all this, the mother is far more passive ; the young have to be all the more active.

In the wild pig (Sus scrofa), I saw some of the young, still wholly enveloped in the embryonic sacs, begin to kick and struggle immediately after birth— head or feet first—and often shake themselves like dogs, thus partly freeing themselves from the amnion. The piglets managed to get quite free, and to dry themselves pretty well by rubbing against their mother's bristly flanks, and even climbing over her, as far as the umbilical cord allowed them.

Sometimes a cord happens to break when the mother gets up to change her position after she has farrowed. Often, however, the cord is broken by the active pulling of the young pig. The behaviour of a new-born wild pig often reminded me of a dog chained to his kennel. At first, the piglet can only go as far from the vagina as the cord, about 15 in. long, allows. The baby pig then pulls at this until it gives way.

As far as I can judge, a basically similar state of affairs exists among the hippopotamus family ; conformity in the position for suckling the young has already been mentioned.

Since observations on the birth of wild animals are still only scanty, and because Asdell (1946) says nothing about birth behaviour, it may possibly be of use to include here some records on antelope and wild pig births.

Active Type
Birth of Nilgai Twins, April 24th, 1949
(Boselaphus tragocamelus)

11.5 a.m. Phone call announcing imminent birth of nilgai.

11.11 a.m. When I reach the antelope house, the nilgai in the paddock has just given birth to a calf, which the keeper has already taken inside on a straw mattress. The calf is being vigorously licked by the mother, and the amnion has already been swallowed.

11.13 a.m. Calf already raising its head a little; drying rapidly.

11.15 a.m. Mother stiffens her tail, strains a little, licks vigorously again, eats a little straw.

11.17 a.m. Calf makes first attempt to get up, but topples over. Ears still limp.

11.22 a.m. The calf replaced on its mattress by the keeper, off which it had rolled, " licked away " by the mother so to speak.

11.26 a.m. The calf keeps tumbling over.

11.28 a.m. Mother lies down, gently straining; head of calf 2 appears.

11.29 a.m. Calf emerges as far as hind legs.

11.31 a.m. Calf 2 already raises its head. Mother at once gets up, turns round on the spot and swallows amnion of No. 2.

11.33 a.m. Calf 1 stands up for a moment, but topples over once more.

11.34 a.m. Calf now standing firmly. Mother licks 2 constantly.

11.36 a.m. 1 now takes its first steps, and looks for the teats in the wrong place, at the front legs.

11.39 a.m. 2 struggles hard to stand up.

11.41 a.m. 1 squats with front legs folded.

11.43 a.m. 1 sucks at the front end and wags its tail. Mother continues to lick 2.

11.44 a.m. Mother lies down.

11.45 a.m. 2 stands up for a moment. 1 now stands steadily. Mother continues licking as she lies down.

11.47 a.m. 2 stands steadily for first time.

11.51 a.m. 1 and 2 stand beside the mother as she lies, and are alternately licked.

11.58 a.m. 1 and 2 try to find the teats at the neck of their recumbent mother.

12.00 Eponychia still quite shiny; glutinous, rubbery, not yet dried.

12.3 p.m. Mother eats some straw.

12.5 p.m. Mother gets up. 1 and 2 again suck at front end.

12.6 p.m. Both search hard at the front and seem quite unable to get away from the wrong place. Vigorous tail wagging and prodding of mother's belly. Occasionally take small tufts of hair between their lips and suck, searching at random.

12.7 p.m. 1 has at last found the teat. 2 stubbornly searches at the wrong place, then walks away. Again tries the front, then stumbles past the flanks and reaches the rear, apparently by chance.

12.8 p.m. 2 at last seizes a teat as well, and sucks.

The placenta appeared at 4 p.m. and was swallowed. The twins are both male.

    1. Weight 16½ lb. Height at withers, 21 in.
    2. ,, 15¼ ,,   ,,   ,,   ,,   20½ in.

Apart from the highly significant search for the teats in the wrong place, often noticed with many other hoofed animals, two further facts about their behaviour are worthy of note. It frequently happens that ungulates in labour nibble at straw, not food. This gives the impression of being a typical substitute, or displaced, movement (in the sense used by Tinbergen, 1940, p. 83).

In addition, we draw attention to the tail-wagging, which can be seen even in year-old Nilgai calves when sucking. This is probably an accompanying reflex phenomenon, which if I am not mistaken is typical of all Bovidae, especially caprinae, ovinae and antelopinae. In most Tourinae, in the narrower sense, it is usually very much slower, or simply consists of a slight raising of the tail, while in goats or sheep it practically vibrates. It seems quite lacking in the Perissodactyla.

Tail-wagging may be seen during artificial bottle rearing. To inexperienced keepers of animals, it is a sign that the baby animal concerned is getting its milk properly. In actual fact, this movement is no guarantee at all that the milk is being taken, as it also occurs when no milk is there, as for instance in nilgai calves sucking at the wrong place.

Passive Type
Birth of Wild Pigs on March 20th, 1945 (*Sus scrofa*)

7.00 a.m. On inspection, keeper sees no signs of imminent birth. True, none has been born so far, yet shortly after, No. 1 appears.

8.00 a.m. The keeper has only just found me, and I watch No. 3 emerge, hind legs first. The sow shivers, keeps getting up and rubbing her snout in the trench.

8.23 a.m. The mother gets up, turns round; this tears the umbilical cord of No. 3, who is still wet and partly covered with the amnion. 1 and 2 already sucking.

8.24 a.m. The boar comes up, sniffs the sow, and at once defaecates close by.

8.29 a.m. No. 4 appears head first, still partly wrapped in its amnion.

8.30 a.m. Mother gets up and in doing so tears the umbilical cord. This one is not licked either. Mother turns round and lies down again. Vigorous wagging of the mother's tail precedes each birth, but she makes no sound.

8.41 a.m. Mother gets up after all four have sucked her. One has a piece of the cord about a foot long dangling from it.

8.43 a.m. Mother lies down again on the other side. Keeps shuddering and quivering.

8.46 a.m. No. 5 appears, breech first. In this case, as in all breech presentations in wild pigs, the hind legs emerge first, the tail held not over the back but with its tip between the legs. The wet piglet shakes itself like a dog coming out of the water, and then remains like the others, linked to its mother at about eighteen inches distance.

8.49 a.m. Mother struggles up to a sitting position, turning over from left to right, thus breaking the cord. Some births take place with sow on her right side, some on the left.

8.53 a.m. All the piglets busy sucking, or trying to suck.

8.55 a.m. All five sucking vigorously. If one of the litter is trodden on or squeezed, it squeals loudly, and then the mother changes her position; squealing is thus a very effective safety device.

9.00 a.m. No. 6 appears head foremost.

9.2 a.m. Mother gets up, severing the cord as she does. Leaves the trench with the six first born, and goes for a drink.

9.5 a.m. Mother lies down again in the trench and allows the piglets to suck her.

9.7 a.m. No. 6 looks in wrong place for teat, namely near the snout. Is still partly enclosed in its amnion.

9.8 a.m. Mother gets up, then lies down on her left side.

9.9 a.m. No. 7 born, head foremost.

9.11 a.m. The new-born baby pig wriggles along its mother's body.

9.18 a.m. Mother stands up a short while, then lies down again.

9.21 a.m. No. 7 goes round the back of its mother, dragging the umbilical cord along, and breaks it off at about twelve inches.

9.27 a.m. No. 8 appears head first, mother lying on her right; is practically shot out. After its head appeared, a short pause; then a sort of small explosion. It crawls over its mother's left hind leg to join the rest of the litter, at the same time severing its cord, without the mother getting up.

9.34 a.m. Mother still shivers and shakes.

9.38 a.m. No. 9 appears breech first, the mother lying on her right; at first is tied to her by the cord, then, as it trots over to the others, it severs its cord at the same time. Mother remains lying down and trembling.

9.49 a.m. Mother gets up and goes for another drink. Leaves the nine young pigs, which can all stand now, by themselves in the trench. On her return, lies down on her left side and roots about, without devouring any of the pieces of umbilical cord that are scattered about, or licking any of her litter.

9.53 a.m. All the young pigs now sucking the still trembling mother.

9.55 a.m. Mother restless; turns on to her other side.

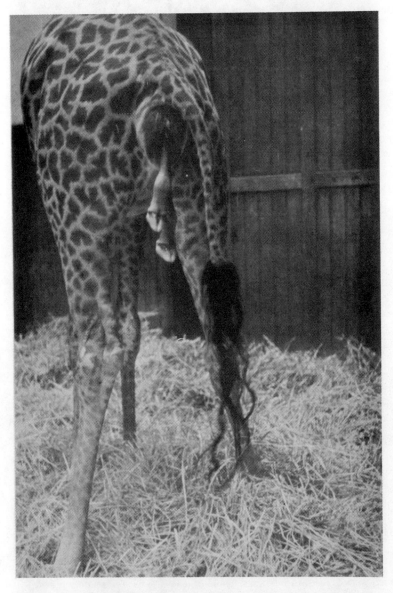

Figure 11a.—Birth of a giraffe. The forelegs have just appeared, the hoofs still encased in the shiny jelly-like skin (eponychia) which protects the mother's birth-canal.

Figure 11b.—*During the birth, which usually occurs standing up, the giraffe occasionally swings its neck in great arcs.*

Figure 11c.—*The longest-legged animal on earth, the giraffe, has the characteristic of giving birth in the standing position, so that the calf literally tumbles into the world from a considerable height, as the illustrations show.*

9.58 a.m. First portion of the placenta, about 10 in. long, appears.
10.8 a.m. Mother somewhat restive; more placenta expelled. One piglet climbs over its mother.
10.13 a.m. Gets up again for a few moments; more placenta.
10.14 a.m. Up again; placenta now hangs down to the ground; grunts; restless.
10.18 a.m. Stands up grunting: expels about eighteen inches of placenta, which is not devoured.

This restlessness continues with interruptions until about 11.30, when a further considerable quantity of the placenta is expelled. We let her lie on purpose and do not return until 5.30 p.m., when the mother and litter of nine have moved to a fresh place.

In contrast to most ruminants, licking of the young and greedy devouring of the placenta are completely absent.

Special factors regarding the mother–child relationship are present in marsupials, yet, as far as giant kangaroos for instance are concerned, they have not been fully clarified. The familiar problem as to how the new-born kangaroo manages to get into its mother's pouch is still open to dispute. Until now, this decisive phase in the mother–child relationship has been observed by very few eye-witnesses, and their evidence is conflicting. As we were able by a stroke of luck, to watch a live new-born giant grey kangaroo on December 29th, 1944, at Basle Zoo, as well as an actual birth on February 12th, 1949, I should like to refer briefly to this problem of animal sociology. Only the giant kangaroo is under discussion.

Today there are four different theories as to how the new-born kangaroo reaches the pouch, and each has the support of different eye-witnesses.

(1) The baby kangaroo actively climbs from the vagina to the pouch without help from the mother, where it firmly attaches itself by suction to a teat. Incidentally, this is the view put forward in the latest edition of Ellis Troughton's excellent work (1948).

(2) The baby kangaroo makes its way quite independently along a track previously licked clean by the mother in the belly fur, from the vagina to the pouch. As H. Dathe (1934, p. 223) reported, this method was observed at Halle Zoo for *Macropus rufus* by F. Schaaf in 1933.

(3) The newly born kangaroo is transferred to the pouch in the mother's paw. Observed with *Macropus rufus* at Leipzig Zoo (K. M. Schneider, 1944, p. 47).

(4) The baby, held between the mother's lips, is transferred to the pouch. This method was confirmed on February 12th, 1949, for the giant grey kangaroo. A similar observation was made before by Le Souef and others (*cf.* Schneider). I am obliged to A. K. Minchin for his interesting letter (September 21st, 1950), in which he informs me that in 1922 he had watched a South Australian Rock kangaroo, and in 1950 a red kangaroo, and saw the mother transfer the baby between her lips.

K. M. Schneider rightly assumes that the transference of the baby to the pouch seems to offer several possibilities, as the different accounts cannot be doubted. It might only be conceivable that the various methods are to be attributed to different species, which are by no means easy to distinguish. The Basle observation (by keeper E. Glücki), in any case, is the first for 30 years to show that the transfer might occur with the help of the lips, which had long been doubted. Troughton's view, that all new-born marsupials find their way to the pouch independently, is thus no longer tenable.

Apart from the fact that the baby, fixed to the teat like a patent fastener, is able to breathe and drink simultaneously, thanks to a special arrangement of the breathing passages, and that the milk is squirted into its mouth by rhythmic contractions of the musculus compressor mammae by the mother, less notice has been given than it deserves to another peculiarity of the mother–child relationship, namely the closing of the pouch.

In this connection, the function of the sphincter marsupii is very important. In her behaviour, the mother kangaroo can be compared to a mouth-breeding cyclid. In the face of danger, it takes the baby into its protecting pouch; the young kangaroo—like the little mouth-breeding cyclids—does not just run away from danger, but into the maternal pouch. In conditions of danger, it may be critically important that the pouch should be open wide enough to allow the quickest possible entry or "jump in".

This sphincter marsupii is usually described as a sort of rubber ring, out of which the baby kangaroo can poke its head from inside and which then contracts around its neck. Moreover, its elasticity is changeable at will, and controlled by the mother. When the baby has to jump quickly back into the pouch to escape danger, the mother anticipates it by opening the pouch. The sphincter muscle is striated, and from the point of view of comparative anatomy, derives from the abdominal wall; I have been unable to find any description of the antagonist muscles. In the zoo we have simply observed the active and as it were, inviting, opening of the pouch, and the delicate synchronization between maternal and infant behaviour.

In marsupials, special temporary relations between mother and child may occur to a striking degree, as is shown by A. K. Minchin's study (1937) of the koala (*Phascolarctus cinereus*). According to him, the baby marsupial is fed at a certain age by the mother per rectum. The opening of the koala's pouch, as is well known, faces rearwards, not to the front. During the change from a diet of pure milk to eucalyptus leaves, the mother, every other day or so between 3 and 4 p.m. for about a month, produces a peptonized pap of chewed eucalyptus leaves, completely different from the excrement. The baby koala thus only needs to stretch its head a little out of the pouch, to receive this transitional food from the rectum.

Little is yet known about the development of the young kangaroo's behaviour during the time it is in the pouch. Richard Müller of Wuppertal Zoo reported an interesting case to the Zoo Directors' Conference at Rome in September, 1952. A young giant kangaroo had to be artificially reared because its mother had died. Eventually, after leaving the artificial pouch, it followed its lady attendant about everywhere. This seems to be the first case of marsupial's "imprinting" ("prägung") on a human being, insofar as Lorenz's term is permissible here.

In Basle Zoo, from 1945–50, we have often been able to watch the first movements of young grey giant kangaroos through the pouch wall, and to continue observations until the young kangaroo looked out of the pouch for the first time, and then one day left it for the first time. In view of the scarcity of such instances, it may be of value to give an extract from the account of two consecutive births:

No. 1. Young Female

May 23, 1945	Mother was mated.
July 26	Baby's first movements seen through the pouch wall.
July 30	Judging from its movements, the baby is about the size of a mouse.
December 4	First the tail, then the head put out of the pouch; very little hair yet.
January 6, 1946	Head and arms out of pouch for first time.
January 12	Eats some hay from out of the pouch.
January 31	Comes completely out of pouch for first time. Stands rather unsteadily and shakily. Tail still remarkably thin; not yet the muscular support of the adult.
March 21	Hops on its own for first time, over door-step between the inner and outer enclosure. When disturbed, invariably jumps back into pouch head first.
April 8	Seen for last time in mother's pouch.
May 19	Still feeds from its mother occasionally, only popping its head to do so into the pouch.
June 10	Still suckled now and then. Half grown.
July 14	Still sucking, although its mother now carries another baby (No. 2) in her pouch.
August 20	Although not yet weaned, was mated for the first time by her father.
December 4	The same observed again.
February 5, 1947	Became a mother; baby definitely in pouch.

The following data about the younger brother of this young kangaroo, already a mother, may be of interest. The mother of the female baby just described was mated on March 3rd, 1946, while she was still carrying it in her pouch. During the father's violent advances, the baby in the pouch was in danger of serious injury, so we sometimes had to shut him up. On April 7th, the mother was mated again.

No. 2. Young Male

April 7, 1946	Mother was mated.
June 17	Young confirmed in pouch. Mother cleans her pouch daily, usually after the older female (No. 1) has been suckled. This free teat is 1½ in. to 2 in. long, the colour and thickness of an earthworm.
November 9	The keeper can see the rather wet baby in the moist pouch; it has a scant coat of hair.
November 24	Looks out of pouch for first time; head and ears out; up comes No. 1, pushes No. 2 back into pouch, and sucks. No. 1's eyes are still opaque.
November 28	Pokes its head right out of pouch for first time. Already quite well covered with hair. Eyes clear.
December 26	Puts head and both arms out for first time.
January 4, 1947	Eats some hay while still in pouch.
January 8	Leaves pouch for first time. Twice the mother replaces the baby that struggles to get out of the pouch, using her snout and paws.
March 1	Sucks now while out of pouch.
March 13	No longer returns to pouch.
March 15	Sucks the mother for the first time at the same time as its elder sister (No. 1) from outside the pouch.

These data, taken together, give the following comparison in the development of two offspring of the same mother:

	No. 1 (♀)	No. 2 (♂)
Mother mated	May 23, 1945	April 7, 1946
Exact date of birth	?	?
First known to be in pouch	July 26 ,,	June 17 ,,
First emergence of head from pouch	Dec. 4 ,,	Nov. 24 ,,
First complete emergence from pouch	Jan. 31, 1946	Jan. 8, 1947
Last time in pouch	Apr. 8 ,,	Mar. 13 ,,

99

How much we may infer from these zoo records about conditions in nature is not easy to say. Perhaps hypersexuality conditioned by captivity (Hediger, 1951) must, as in analogous cases, be taken into account *i.e.* a prolongation of the lactation period, precocious maturity, and rapid sequence of births. Most probably, the reproductive tempo of kangaroos in natural surroundings is considerably less than under zoo conditions. The baby depends on its mother for nearly ten months. If one reckons between 38 and 42 days for the duration of pregnancy, then there is nearly a year's dependence on the mother.

These few examples show that in mammals, too, certain kinds of mother–child relationship may be established, of importance from the point of view of heredity. With regard to conditions among the cetacea, E. J. Slijper (1949) published a summary of birth behaviour habits hitherto recorded for mammals. Some details about peculiarities in the births of various animals are contained in the book by K. de Snoo (1942) on the problems of the beginnings of mankind in the light of comparative obstetrics.

Among the most important problems of comparative psychology is that of the proportion of the innate to the acquired. How much in man's behaviour is so to speak present automatically, through inheritance, and how much conditioned by the effects of human surroundings, by tradition. The instances of men brought up by animals in an animal milieu, for example the much-discussed report on the wolf children of Midnapore (Singh and Zingg, 1942), rare in themselves and thus difficult or impossible to verify, would seem to show that very little specifically human develops under such conditions. Such wolf children are idiots; they lack human speech and do not walk upright. But all these reports are too inaccurate and unreliable for any clear conclusions to be drawn from them. On the other hand, we should not be justified in producing wolf children experimentally by handing over human babies, under scientific controls, to wolves or other animals to bring up.

In view of these facts, Professor Lightner Witmer in 1909 already drew attention to the fact that the crucial experiment, the obverse of the impossible wolf-children experiment, was quite possible, and even desirable, in the interests of science; that is, the bringing up of animals in human surroundings. This would be bound to show how much of the purely animal side developed, and how much—or little—an animal can be humanized. Obviously, the anthropoid apes, those animals that most closely resemble mankind, seemed most suited for this experiment.

There were soon in fact numerous experimenters who took this theoretical challenge seriously, and brought up anthropoid apes—chimpanzees—just like human children, even together with human babies, and under identical conditions. The most valuable experiment of this nature was undertaken in 1931 by the American scientists, Mr. W. N. and Mrs. L. A. Kellogg. The Kelloggs adopted a seven and a half months old female chimpanzee called Gua, and brought it up with their own baby, Donald, two and a half months younger, under absolutely identical conditions. Gua and Donald had similar clothes, similar beds, similar attention, similar food, *etc.* Both have been written about at great length, forming the highly-documented subjects of a 341-page book (Kellogg, 1933). Their picture appears on the title page, hand in hand, a really remarkable couple.

A surprising amount of most interesting facts are contained in this account, which until recently was accepted as the standard work on that particular subject. To anthropologists, psychologists, doctors, philosophers and zoologists, it represented an incomparable source of material, presented for the first time. The work has only one drawback, or rather suffers, understandably enough, from one limitation, namely, that this unique example of co-education lasted only nine months. Then the " chimpanzee girl " was carefully broken of its habits, and taken back to Yale University Anthropoid Ape Farm at Orange Park, Florida, whence it had originally come. No harmful effects on the development of its human partner, Donald, were noticed.

The demand for a longer and more comprehensive experiment remained. Two young married American psychologists, Keith and Cathy Hayes, remembering this unfulfilled scientific wish, in 1947, formally adopted a new-born female chimpanzee—Viki Hayes—on her third day of life, from the same place, the Yerkes Laboratory of Primate Biology. It is practically impossible for an outsider to realize what this must have meant, and still means, in the way of personal sacrifice. Every parent knows only too well how tied one is to a baby in the truest sense ; but the tyranny resulting from having a baby chimpanzee to look after exceeds anything else, especially from the time it begins to leave its bed, and clamber about the kitchen, into the cupboards, and on to one's writing desk.

From the outset, Viki was determined to allow no baby sitters—she insisted on her foster parents' constant attention. This meant that for years they were never able to go out together, to say nothing of enjoying a well-earned holiday in each other's company. In spite of the extraordinary strain and constant tension, Cathy Hayes somehow found the time and energy to write a biography of Viki's first three years, not in the form of a dull report, but as a most vivid account, and containing as many amusing incidents as scientifically valuable facts (C. Hayes, 1951). It is not hard to predict that this three and a half dollar book will become a best seller, and will certainly be translated into many languages.

The publication of this unusual book does not mean the end of the Viki experiment, but only a sort of introduction to it. During my visit to Florida in 1951 I had the opportunity of paying Viki a visit at her foster parents' house, when she was over four years old.

Viki is just as pleased as human children when visitors arrive, even when they have forgotten to bring a small present. I was given a lively reception by Viki in the beautiful house near Orange Park, set in the most wonderful green surroundings. She had just had a bath, was wearing clean rompers, and was far and away the daintiest little chimpanzee I had ever met.

As we were sitting in the discreetly wired-in sitting-room, Viki sat confidentially and inquisitively on my lap, examining my tie as carefully as my face, or my wristlet-watch. Among other things, I discussed with her foster-parents the desire and the ability among apes to mimic, which I had not thought to be highly developed until that moment. Doctor Hayes happened to mention that Viki at once put on any shoes she managed to get hold of, so I pulled one shoe off and put it, as a sort of test, beside my armchair. Viki immediately slipped it on, and to my relief clumped around the table with it, delighted. But before she had completed her first circuit, she espied a tiny hole in my sock. Intrigued, she at once ran up on all fours, examined it, and finally threatened to start playing with my toe so that a defensive gesture became imperative. My host suggested that I cover the distracting hole with my other foot and, thanks to further diversions we were able to continue our conversation a little longer.

But soon Viki took me by the hand, exactly like a child, and pulled me with such an unmistakable and irresistible expression, that I had to follow her into the playroom next door, where she kept her dolls, animal toys and picture books. She too adores turning over the pages of old magazines. One day she found an advertisement in one, showing a wrist watch natural size. Then something surprising happened ; she pressed her great ear to it and listened for the ticking, which she had always found so fascinating in real watches. Viki thus showed how she recognized pictures, and therefore accomplished something of which hundreds of thousands of the species *Homo sapiens* are incapable. Explorers who have taken photos of natives of primitive races in Africa, South America, New Guinea or India, have always noticed that these people could not recognize a man's picture without special help. I too was often aware of this fact in the Pacific islands.

In Viki's case, this recognition of pictures was most carefully tested, incidentally with her gift for imitating. Her foster-parents taught her to mimic all sorts of actions at the command " Do this ".

If, for example, someone shows her how to hold both ears with two hands, she imitates the movement. The surprising thing is that she can be made to imitate such movements by being shown photographs of them.

This, however, is just one of the countless really astonishing observations made by the Hayes on their foster-child. We confine ourselves to saying that Viki has even learnt to say a few words, *e.g.* papa and mama. In her case, they were not satisfied with discovering the limits of normal capabilities, but took the trouble to develop to the uttermost, and with special aids, all the latent capacities of this species under human conditions. Anybody who is the least interested in biological problems will not be able to ignore Cathy Hayes' book, for this continued experiment represents a novel and striking advance in scientific research. Unfortunately, Viki died on May 11, 1954.

8

WILD AND DOMESTIC ANIMALS

GEORGES CUVIER (1769–1832), the famous French naturalist, was convinced that animals, even domestic ones, were fixed and unchangeable. As they had been created in the beginning, so they would remain for ever. Since then it has become evident that this theory is no longer valid. Domestic animals show how man's influence has changed them through the centuries and millennia by breeding, how they have developed from wild animals to domestic creatures, often a mere caricature of their wild ancestors. One has only to think of the unwieldy farmyard pig, a waddling roll of fat, and compare it with the nimble wild pig ; or of the fragile pekinese, compared with the wolf.

In many cases we can follow the development of domestic animals step-by-step from the wild ancestral forms, and can thus watch the increasing transformation of the wild type. Hans Nachtsheim (1949) has shown that the origins of the domesticated rabbit, of which scores of breeds are known, date back to Roman times. In the time of Caesar, young captive hares— both common and mountain—were kept in walled enclosures, the so-called leporaria. Soon, however, rabbits were preferred because they increased by breeding. The Roman scholar, Varro (116–27 B.C.), left a detailed description of these rabbit pens, which lasted with hardly a change until the Middle Ages, as we learn from a manuscript of 1393. It seems that it was then a sport of noble ladies to shoot the harmless but tasty game in the leporaria with bows and arrows. In Switzerland, Germany and England there were no rabbits, wild or tame, at the beginning of the Middle Ages. The classic land for rabbits was Spain. A coloured woodcut of 1423, according to Nachtsheim, is the first picture of a wild rabbit from Germany. It is leaving its burrow in its very typical attitude at the feet of St. Christopher of Buxheim, the chief figure in the woodcut.

Once the breeding of rabbits had caught on in Central Europe, the first colour varieties appeared comparatively soon; for instance the grey in 1631 in England. In the nineteenth century the so-called Russian rabbit was bred, a colour type whose extremities, ears, nose, paws and tail, are black. About 1900 some dozen different varieties were known, since when this number has increased. Thus the whole of this large group of breeds can be traced back to its source.

One of the most recently domesticated animals, far newer than the rabbit, is the golden hamster which has of late found many enthusiasts and has become a great pet, replacing the white mouse. In England there are even golden hamster breeding clubs, with their own periodicals. In Europe and America today there are tens of thousands of these clean, pretty and interesting rodents, where they are being looked after not only by enthusiasts, but in almost every biological laboratory as well, and in 1950 the first colour mutations appeared. Thus they can properly be described as domestic animals.

The history of their domestication is very well known and very short. It has been described as follows in H. W. Reynold's short report (1950). The golden hamster it is true had been bred once before in England in the previous century, but the whole brood completely died out. In 1930, Professor Ahorni caught a female with twelve young, when on a scientific expedition near Aleppo in Syria.

From these thirteen specimens, all the countless thousands of golden hamsters kept as pets and laboratory animals today are descended. This handful of hamsters arrived at the Biological Institute of the Hebrew University of Jerusalem. In the following year, 1931, Dr. S. Adler presented two pairs to Dr. Edward Hindle of London, who first perfected the proper method of breeding from this most interesting little animal once so rare. It will interest Swiss enthusiasts to know that the first specimens came to us in 1946 direct from Dr. Hindle.

An interesting biological point about the history of the golden hamster is that its swift spread has not so far been accompanied by any harmful effects, in spite of all imaginable risks from inbreeding; in this it is similar to the increase of the musk-rat and to the avalanche-like spread of rabbits in Australia. All these are cases of breeding from very few pairs; that is, of continued in-breeding.

The reason for dwelling so long on the domestication of the rabbit and the golden hamster is because I wished to show from both these examples how we can watch the emergence or development of new species or races of animals from the very start. This is of course not true for all domestic animals, but only for the latest ones, including the budgerigar, the platinum fox, the mink and a few more. With these we can state the exact date of the appearance of conspicuous hereditary deviations, that is striking mutations from the original natural state.

It is a very different matter with the so-called older or primary domestic animals, whose origins are hidden in remote antiquity, and whose wild ancestral forms died out in historic or even pre-historic days; for example, the aurochs, the quaternary horse, *etc.* The dog, as everyone knows, is the oldest domestic animal, its domestication dating back to the middle Stone Age, *i.e.* to the period between the 16th and 6th millenary B.C. The dog, *i.e.* the domesticated wolf, was the first creature with which man got on to intimate terms, or that got on to intimate terms with him, and which in the course of thousands of years became uniquely intensified. No other animal stands in such intimate psychological union with man as the dog, which has almost become his master's thought-reader, reacting to his faintest changes of expression or mood.

Next comes the horse, to which man may stand in close psychological connection. It became domesticated in the late Stone Age, *i.e.* between the 6th and 2nd millenary B.C. Cat lovers, who claim equally intimate relationships with their pets, may be right, though their domestication took place only about 4,000 years ago.

The fact that the dog is by far the oldest domestic animal does not necessarily imply that its wild ancestor, the wolf, was the first wild animal to be kept in captivity by man. There is considerable evidence that in Alpine districts in prehistoric Europe, the powerful cave bear was kept in rough cave grottoes by the primitive men of those days. And since the discovery of fossil giant sloth bones in South America, many scientists have become convinced that those long-extinct monsters, varying in size from a cow to an elephant, were kept in captivity by primitive men. One of these giant animals was called *Grypotherium domesticum* by its discoverer, the Swiss palaeontologist Professor Santiago Roth, a former saddler's apprentice, because he was sure that it had been domesticated, or at least kept in captivity, by man.

The exciting thing about these giant sloths is the fact that not only have their fossil bones been found, but pieces of skin and hair as well, so that in the nineties of the last century, a few optimists thought it might still be possible for some living specimens to be found in the countless caves of the coastal districts of Southern Patagonia. Expeditions for this purpose were even sent out, without result unfortunately, as Brehm in his *Life of Animals* somewhat spitefully observed.

Nevertheless, certain scientists believe that man kept these giant sloths regularly in captivity in Southern Patagonia. This fact was taken into account in the reorganization of the Department of Mammals in the London Natural History Museum after the war. There, in 1951, I was able to admire a piece of the skin of *Mylodon listai* one of these extinct animals, and to read with interest the appropriate label: " Skin of *Mylodon listai*. This animal is one of the extinct giant ground sloths. It was a native of South America and a contemporary of man. This piece of skin was found in Southern Patagonia in a cave where the animals had been kept in *semi-domestication*. Part of the cave had been walled off and in one corner was a pile of hay."

It is not my purpose, however, to go into the history of domestic animals or of species partially domesticated by man, however attractive this might undoubtedly prove. My concern is principally to show that man's efforts to tame wild animals, to keep them in captivity, and to domesticate them, are as a matter of fact as old as man himself. At all times, and in all parts of the world, man has always striven to do this to animals, and still continues to do so till this day.

New animals that did not occur in nature have been evolved through this universal effort, which has gone on since the Stone Age. Let us consider for a moment what the hypotheses, essentials and results of this experiment are. One thing may be expressed on the basis of a simple numerical and a geographical consideration: compared with the enormous numbers of wild animal species, domestic animals comprise a trifling minority, less than one in a thousand. If we think of the vertebrates alone, the comparative total is rather larger, lying between one and two per thousand.

We need not bother about actual figures; the important thing to realize is the insignificant number of domestic animals, forming only a minute section of the enormous reservoir of wild animals. What then were the decisive points for choosing them ? This choice was certainly not random, but depended on a very definite coincidence of factors in animal and human psychology. On the part both of man and animal, a readiness to domesticate and be domesticated was necessary. A two-fold pre-requisite is clearly implicit in domestication, one on the animal's behalf, and the other, no less decisive, on man's. The possibility of the creation of a domestic animal can only result from a fusion, or union, of both.

On the animal's side, a sort of latent readiness for contact with man and for sharing his life, are especially necessary, as well as a certain inclination for human buildings and constructions (" technophilia "). It is not likely that markedly technophobe animals, the so-called avoiders of civilization, should react positively to man's approaches, since they avoid him and all his influences as far as possible.

Another pre-requisite on the animal's part is its capacity for being bred in captivity, as domestication, in contrast to individual taming, depends upon inbred influences working through generation after generation. This is probably the reason why the cheetah, for example, cannot be domesticated, although it is otherwise to a considerable extent favourably disposed. Up till the present, no zoo in the world has managed to breed these lovely, elegant spotted cats.

In an earlier work (1938), I put forward the conjecture that special importance is attached to biological rank—especially among large animals—in so far as many wild ancestors of our contemporary domestic animals have taken up, or are taking up, a subordinate position in the biological hierarchy. Biological α-species, such as bison or ibex, unaccustomed by nature to be subordinated to any other creature, do not seem to have any disposition to become domesticated. This special kind of pre-adaptation is all the more pronounced, however, in biologically inferior species and races. Thus, for example, it was not the powerful bison that was domesticated, but the aurochs, of much smaller stature; it was not the large wolf, but a smaller race, not the huge wild horse of the steppes (Antonius), but the little grey forest one, *etc.* We often find greater tameability in biologically inferior species than in α-species.

The line of thought referred to here has unfortunately been misinterpreted or misunderstood in certain quarters. Thus, for instance, Krumbiegel (1947, p. 31) reads into it the " so-called theory of inferiority ", according to which there are said to be unapproachable master animals, and " inferior " creatures, suitable for developing closer relations with man. O. Antonius (1938, p. 296) even deduced from my working hypothesis for the animal concerned characteristic " inferiority complexes ", and applied them solely to the domestic cat. No kind of inferiority attaches in itself to biologically subordinate species : why for instance should the chamois be inferior to the ibex ? Individual species or races merely assume different functions within their biocoenosis (or association of different animals that live in a certain area) as defined on page 67. G. Steinbacher (1938) confused the biological hierarchy with the totally different predator–prey relationship.

Previously, it was believed that the social way of life, living in packs, herds or swarms, was the decisive criterion for domestication. Today, however, it is clear that sociability by itself is no guarantee of predisposition to domestication; the domestic cat in fact descended from a distinctive solitary species, the Egyptian *Felis ocreata*. If all other pre-requisites are satisfied, sociability may, of course, have a beneficial and strengthening effect, as we see in the case of the dog.

A special disposition was needed on man's side too; a gift for handling animals, for looking after them. The lack of this quality is even more striking perhaps than its presence. Thus African natives are less gifted in this respect than, for example, Indians or American Indians. In spite of its enormous wealth of animal life, the great African continent has, significantly enough, produced hardly any domestic animals—an extremely surprising fact, really, and one which still has its effects today. The natives of Africa are obviously no good at looking after animals. Anyone who has had anything to do with natives and with animals in Africa will have had painful confirmation of this.

Apart from the donkey, only the cat and the guinea fowl have been domesticated in Africa. It is possible that Asiatic influence was at work in the case of the donkey, and as for cats and guinea fowl, they practically attached themselves obtrusively to man.

Their predisposition was an unusual one and the proof of this is to be seen in the behaviour of the wild representatives of today. The famous African explorer Georg Schweinfurth (1836–1926) had already pointed out the surprising speed with which young Egyptian yellow cats grow tame and get used to houses; they seemed to him to be made for domestication. Professor Otto Antonius confirmed this, after many years' experience with animals in Schönbrunn, Vienna. When caught young and reared by natives the wild guinea fowl in Africa stay in the vicinity of the village and I myself was able to see what nuisances they are and how hard it was to get rid of them. Even while in the temporary company of completely wild guinea fowl, one that has been so reared can never bring itself to join up with them.

The failure of the African native to assume any relationship with wild animals, other than that of hunting them, is a constant surprise, considering the abundance of antelopes, zebras, buffaloes, *etc.* Only the ancient Egyptians were outstanding as keepers and breeders of animals. They partially domesticated several birds (Nile goose, cranes) and mammals (antelopes), but later their attempts at breeding disappeared completely with a few exceptions. The breeding of Watussi cattle, *i.e.* of Inyambo for ritual purposes, to which I shall presently return, did not originate in the present territory, but has come down with many modifications from ancient Egypt.

About the turn of the present century, white men again began the experimental domestication of eland and Grevy's zebra. In any case it was not completed, partly due to the rapid mechanization of transport, partly to the increasingly successful control of tropical epidemics, in particular of nagana, to which horses and cattle are highly susceptible.

In the classic elephant lands of Africa in Belgian Congo, where even today between ten thousand and twenty thousand elephants are killed annually the natives have never tried the experiment of taming an elephant. When in 1880 the missionaries of Sainte Anne-de-Fernard Vaz in the Gaboon succeeded in doing so, the natives came from miles around to see this miracle, which became to some extent the starting point for the present Station de Domestication des Eléphants (S.D.E.) in the Congo.

It must be mentioned in passing that this unique state elephant enterprise is not an establishment for domestication, but only for taming, as only wild elephants are broken in, and no elephants are bred there. As we have pointed out, it takes generations to produce domestication. In a report I had to prepare in 1950 for the Institut des Parcs Nationaux du Congo Belge on this elephant station, I came to the conclusion that the African elephant could not be domesticated, and could never be of material economic assistance to man like the Indian elephant, simply because it could no longer compete successfully with motors and machines. A lorry or tractor is much easier to handle than a full-grown African elephant, quite apart from the problems of feeding. The African elephant has an abnormal digestive capacity, with very poor assimilation. It must consume so much bulky and moisture-containing food that it has little time left for working after feeding.

Now that we have become familiar with some of the essential pre-requisites of domestication, some of its principal effects must now be mentioned. Here the essence of domestication, the difference between wild and domestic animals, becomes self-apparent. Since representatives of either type are kept side by side in zoos; zebras and ponies, guanacos and llamas, mouflon and domestic sheep, pheasants and domestic fowls, the differences between them are clearly seen in the zoo.

By far the most important difference between wild and domestic animals lies, as we have said, in flight tendency and enemy avoidance.

Let us take the horse, as a concrete example of zoo practice. Its hoofs and coat are cleaned daily. The animals stand quietly in their stables, and raise each leg for hoof-cleaning automatically, corresponding to a symbolic invitation. If the same thing were attempted with zebras—that is, with wild horses, there would be some surprises in store. Normal zoo zebras only tolerate entry into their stables with certain reservations, and would not permit inspection of their hooves under any circumstances. Because of its flight tendency, to a considerable extent reduced through taming, the wild horse allows no contact of that kind. Even trivial attention to the hooves, often due to uneven wear from lack of enough exercise, must be performed with anaesthetics for the protection of the animal and the man concerned.

The modern domestic horse itself has in the course of countless generations not quite got rid of the original flight tendency of the wild horse, although it has been in man's service for a good five thousand years. This does not happen by chance. The tropical wild horses, zebras, live on open grass lands, and are the favourite victims of the great cats. In face of these enemies there is no other flight reaction than the quickest possible dash away over great distances. And this may be necessary at any moment.

Other animals can escape from their enemies by climbing, flying, digging, or "freezing". Wild horses and other hoofed animals can only run away. This characteristic feature is the focal point, so to speak, on which all the other psychic and somatic properties of the horse are centred. True, it has been overlaid by thousands of years of domestication, but never entirely eliminated. This can be seen in daily life, sometimes with drastic or even tragic results. In Switzerland alone, some fifteen lives are lost every year on the average through the latent flight tendency of horses being set off by some incident—in other words, the horse bolts.

We then read this sort of thing in the newspapers: (Case I, 1950) "Rebstein (St. Gallen), 8. May. Last Saturday morning a serious accident occurred in Rebstein. Two horses that were being harnessed by their driver suddenly shied and bolted off together with the cart, knocking over two children playing in the street. One was rescued, but three-year-old S.R. was run over by them, and received serious injuries, from which he died soon after."

A further example : (Case II, 1950) "Lengnau (Berne), 29 June. Last Tuesday in Lengnau 66 year old H.G., a widower, was knocked down by the horse that he was leading by the reins, when it suddenly bolted. He fell under the wheels and received fatal injuries."

An analysis by an animal psychologist of similar accidents with horses shows in nearly 100 per cent of the cases an explosive reactivation of the characteristic flight tendency on the part of the horse.

Accidents with cattle, especially with bulls, are in quite another category, and result in the deaths of some 13 people in Switzerland annually. These accidents have no connection with flight tendency, but have in the majority of cases as motifs fights among rivals. Here the previously mentioned assimilation tendency typical of so many domestic animals, is at work. The bull suddenly attacks the man because it sees him as a social, often even as a sexual rival (though seldom as an enemy of another species).

Reports about accidents of this sort usually run like this (Case I, 1950). "Meiringen, 22nd May. 61-year-old J.K. of Schattenhalde near Meiringen was thrown to the ground by a mad bull and so badly gored that he succumbed to his injuries."

Or (Case II, 1946). "Cully, 21st March. A farmer from Bilaz, near Savigny, was attacked by a bull as he was about to take it out to drink. While still in the stall the animal gored the unfortunate man, its horn piercing his heart. It then trampled on him. The body was discovered later on by the son of the deceased."

This all goes to show how closely animal psychology can be connected with practical life; here it could directly affect the prevention of accidents. If one wished to draw practical conclusions from this obvious analysis, there should have been some means of curbing the horse, and preventing it from receiving flight-releasing stimuli. Both of these have long been in use with varying success, in the shape of snaffles and curb-bits for the former, and blinkers for the latter. Naturally, the partial elimination of disturbing visual impressions is no remedy for eliminating the causes of flight.

As for accidents with bulls, these must be avoided by treating the animals exactly as wild bison or buffalo bulls are treated in the zoo. This would, of course, mean all sorts of complications, but there is no doubt that it would also increase safety.

Incidentally, dog bites, which are fortunately seldom fatal, are another matter, and are nearly always due to social causes, because dogs regard men as a social rather than as a sexual rival.

We shall see now how the horse's latent readiness for flight is not always a disadvantage. The stimulating effect of a whip gently waved is based on its importance as a symbolic flight release, and a light rein leading a horse is simply a symbolic one-sided restriction on flight.

A similar contrast to the flight tendency of horses and zebras can be seen for example in wild hares and domesticated rabbits. While the rabbit can be kept in small hutches, and allows itself to be picked up by the scruff of the neck when the hutch is cleaned out, the wild hare needs a special cage, the back-to-back cage with alternative compartments, in which its lightning-quick flight reaction can be curbed, and even to a great extent avoided, while the extremely timid animals can never see their keeper at a distance of less than twelve feet away.

The same thing is found elsewhere. If we enter the enclosure where wild sheep—for instance mouflon, barbary sheep or Himalayan tahrs—are, we at once touch off their flight reaction; but with domestic sheep on the other hand, this happens to a far lesser extent.

With domestic animals, protected by man from their natural enemies, there is no longer any need for the strict uniformity of wild colour in nature, and we soon find, after domestication, wide variations in coloration; *e.g.* white, checked patterns, melanism, flavism, *etc.* This phenomenon is frequently much more pronounced in birds than in mammals, thanks to a combination of pigmental and structural colours.

Of course mutations from the natural coloration also occur in the wild, but variations like these usually lead to increased exposure to enemies, and such cases disappear soon after they first occur, thanks to selection. Uniformity is thus maintained through selection, conditioned by the presence of enemies. A convincing example of this was seen in Basle Zoo. Among the many peacock chicks hatched in the gardens each year in freedom, the white ones disappeared first year after year so regularly that it was thought that white peacocks could not be reared in the zoo. Not until we checked this selection, caused by predators such as owls, stoats, weasels and the like, by catching the newly-hatched white peacock chicks, and rearing them in the safety of a stout cage until they were the size of pheasants, did we succeed in keeping white peacocks in the dangerous freedom of a garden even when protected by a fence. An additional factor is that in addition to their unbiological coloration, albinos are often constitutionally weak.

Colouring apart, there are other characteristic ways of avoiding the enemy which, thanks to domestication, *i.e.* to man's intervention, lose some of their effectiveness, in the sense of rendering the animal less able to survive in natural surroundings, for instance, power of movement and defensive weapons.

In their ability to overcome obstacles, domestic and wild animals differ diametrically. For example, while a chamois or an ibex can clear a twelve foot fence in an emergency, less than half that height is enough to keep domestic goats or sheep in. There is a marked difference in endurance, too. Far less effort is needed by the zoo staff to recapture an escaped domestic animal than a wild one. This is not the place to go into the various causes of this fact—physiological, anatomical or histological—which are familiar in all zoos.

Sometimes man has deliberately aimed through breeding at reducing his domestic animals' powers of locomotion—the short-legged sheep, the strange "fainting goats" and the so-called "creepers" among hens, are extreme examples of this. Exaggerated plumpness, excessive fatness, *etc.*, are further examples of obstacles to flight, *i.e.* of the reduction of powers of movement owing to domesticity.

On the other hand man is able to supplement by secondary means this lack caused by domestication in those domestic animals from which he requires special powers of movement. The most striking illustration of this is the horseshoe. Wild horses, zebras for instance, although sometimes living in very stony country, have naturally such exceptionally hard hooves that they need no protection against wear and tear. One is continually surprised that zebras, whose hooves need constant attention in zoos, at considerable trouble, have in the wild reached such a complete harmony between continual growth, regular wearing down and degree of activity. In captivity, that is in a condition of domestication, the interplay of these three factors is nearly always disturbed, so that remedial measures must be undertaken.

Another class of organs, defensive weapons, has been considerably modified by domestication, though in wild animals living in freedom it plays a decisive part in avoidance of or encounters with enemies. Both the horns of ruminants and the teeth of predatory animals are either weakened or grossly overdeveloped by domestication, as in the case of the giant-horned Watussi cattle.

There is a very close connection between the weakening of the impulse to be constantly on the alert for enemies in the domesticated wild animal, and the reduced effectiveness of those organs that serve to give timely warning of enemies in the dangerous conditions of freedom, namely, the sense organs. This reduced effectiveness is very often coupled with an externally visible deformation and leads to unbiological or useless development. The ear, for example, is a striking illustration of this.

If we watch the working of the external ear of a ruminant, a stag or an antelope, or even a kangaroo, we can easily see how both receptors, often independently, are constantly cocked in the direction of every source of acoustic stimuli. During this ear movement, largely automatic, the left ear may be directed towards the front or the rear, while the right ear is following a moving source of sound in a semi-circular motion.

In domestic animals, this power often becomes completely lost. Between sheep, some with almost complete lack of ears, or at best with only rudimentary ones, to the grotesque hypertrophy of hanging ears in rabbits, bloodhounds, goats and pigs, there are all imaginable degrees. Frequently the external ears are not atrophied as a result of breeding, but have been clipped in the individual concerned. This is not therefore an unbiological mutation.

Since the external organs of sense, particularly the outer ear, are at the same time extremely important organs of expression, their atrophy or grotesque deformation results in a substantial modification of the powers of expression. With few exceptions the domestic animal's powers of facial expression and gesture show a more or less noticeable poverty in comparison with the wild animal's.

This fact is particularly striking if we include in this rapid survey other expressive organs, *e.g.* the tail. The tail is popularly thought of as the barometer of the animal's moods, and is in fact a first class indicator of expression. In dealing with wild animals, the finer shades of their emotional condition may often be gauged from the position and movements of the tail, while in domestic animals this organ is either docked, or deprived of its original function. There are even tail-less or stump-tailed varieties, dogs and pigs with immovable curly tails, sheep with enormously exaggerated tails, or tails that have become shapeless, unwieldy bags of fat.

In domestication there are always deviations from the normal, the useful, the biological and towards excesses in both directions. Thus the horns of wild Bovidae (cattle, antelopes, sheep, goats, *etc.*) are organs precisely adapted to particular functions. H. Bruhin discusses this subject in a thesis (1953).

In domestic conditions, horns, which are so highly perfected in the wild animal, are distorted in both directions; at one extreme in hornless cattle, for example, which are easier and less dangerous to handle, and can be accommodated in greater comfort and numbers in cattle trucks. Nowadays the considerations to be taken into account are largely technical and economic, not biological at any rate. Occasionally, however, there are ancient ritual considerations, at least in Africa, and it might be well to remember this ritual aspect, since in it the roots of nearly all attempts at domestication are to be sought.

At the other extreme, we sometimes see in zoos the so-called Watussi cattle, famous for their enormous horns. The name Watussi is somewhat misleading. These long-horned cattle are found in Ruanda, a mandated territory of Belgian Congo, and lying to the east of it. They are simply an inferior form of the Inyambo, or in other words the Inyambo are the pick of the Watussi. To my knowledge, no real Inyambo have yet been seen in Europe; they are the personal property of the King of Ruanda, the Mwami, and cannot be sold.

Some years ago, I had the opportunity of making the acquaintance of these Inyambo on the spot, when on a visit to the Congo and Ruanda. These magnificent beasts, with their gigantic horns, sometimes with a circumference of eighteen inches at the base, are in fact still purely ritual animals, and are most carefully tended and protected. For each animal, there are two or three black attendants, whose functions are almost those of priests. Under no circumstances is an Inyambo cow killed, milked or bled, as is sometimes stated. The only useful things obtained from it are the dung, used as fuel, and the urine, used in cleansing the body.

As these Inyambo serve no useful purpose, they represent a heavy burden on the colony, especially as they eat up the best parts of the otherwise poor grasslands, on which fat cattle could be kept. Their numbers are continually increasing since they may not be killed, and especially since they have been attended to and vaccinated by state veterinary surgeons for the control of rinderpest. This has proved necessary, otherwise the Inyambo herds would have constituted a continuous source of infection to the surrounding herds of cattle for human use. Thus the paradoxical situation has arisen whereby the government must spend considerable sums to maintain a herd of cattle, which are not of the least practical use.

The importance of the Inyambo is purely ritualistic, and is therefore closely connected with the original significance of all cattle rearing. For one

Figure 12.—Example of clever, helpful interference by a trained wild animal in a human situation : a brown bear (lower left) comes to the help of the trainer and his assistant, who are trying to get an escaping Himalayan bear back into the ring.

Figure 13a.—*Practical use of flight reaction in the ring : by simple infringement of the flight distance, valid for the actual situation by the trainer, the lion can be driven from his 'place' to any desired point in the ring.*

Figure 13b.—*Practical use of the critical reaction in the ring : by infringement of the critical distance, valid for the actual situation, by the trainer or his 'extension' (the whip-lash), i.e., by carefully calculated provocation of defence actions, the animal can be* manoeuvred *to any desired point in the ring.*

reason alone—their scientific importance—this unique herd is worth preserving.

One result of domestication to which we must now turn is the general decay of specific ceremonial, and of the space and time system. Both may be observed in the European bison, those rare wild cattle, of whom only about 120 specimens now exist.

It is the rule among wild animals that mating only takes place in certain definite conditions, chiefly seasonal. Most species of wild animals are only capable of mating, or conception, at definite times, the period of heat. This period is often confined to a few weeks, or even days, in the year. With domestic animals, on the other hand, periods of heat are mostly completely divorced from the seasons, and often spread over the whole year.

We may thus speak of the hypersexualizing effect of domestication, especially since sexual maturity occurs earlier in domestic animals than in their undomesticated relatives, since births follow each other more quickly, *etc.* In contrast to the domestic animal, which is usually not particular, and without more ado mounts a rough imitation with only the remotest resemblance to a normal sexual partner, the wild animal often needs a complicated obligatory mating ceremonial, a well-prepared rite.

This is connected with the fact that among domestic animals the most fantastic cross-breeds are possible, while among wild animals such a thing only happens exceptionally. As long as one partner fails to behave according to the ceremonial traditional to the species, no union is possible.

We had a vivid instance of this in Basle Zoo during the war, when it was not possible to provide our solitary European bison bull with a female of its own species, a bison cow. In despair, we tried to introduce a cow of the brown Swiss variety, on heat, to the then 13-year-old bison bull under all possible precautions. The bull took little notice of the cow of a different species, and it never occurred to him to mount her. Two years later when we succeeded in obtaining a bison cow, the now 15-year-old bull mounted her at once. With domestic animals, sexual partners can be much more varied, much less specific.

Above all the whole reproduction behaviour of the domestic animal is much more relaxed. The roughness and pitilessness of the single acts of behaviour disappear with domestication, from the mating ceremony to birth, suckling and so on. Among wild animals, an extremely powerful selective influence is at work, so that the slightest variation from the norm inevitably leads to exclusion from the breeding community; while the domestic animal has relied on man's help for generations, particularly in the case of births, and without it would in many instances have died out.

In wild animals, especially ungulates of the active type, the first duty of the mother after she has given birth, as we have seen, is to free the baby animal from the amnion, which she does by eating the latter greedily. With domestic animals on the other hand, it may happen that the mother brings into the world a well-developed baby, and then at the last moment lets it suffocate with complete indifference in the unopened amnion. This is quite unthinkable in a wild animal.

Finally, among wild ruminant ungulates, there follows intensive licking of the baby animal, and this dries its coat, empties the bladder and intestine, and gives it a general stimulation, at the same time helping it to stand up for the first time. All these natural methods of behaviour during and after birth may be completely absent in ungulates in a domesticated state, or may be supplemented by man. Thanks to his intervention, even defective mutations may reproduce, which would be quite impossible among wild animals. In this way the domestic animal's dependence on man grows from generation to generation. Artificial insemination, implantation of fertilized eggs, Caesarian section and artificial rearing have for long played a part in the breeding of domestic animals. In this way, it has developed more and more into an abstract, unbiological (*i.e.* unnatural) creature.

The relaxation of specific ceremonial, especially when this leads to complete disappearance of reproduction ceremonial, represents one of those phenomena of domestication that have been exhaustively studied by K. Lorenz (1940). Lorenz rightly speaks in this connection of phenomena of disintegration, and the process of " dedifferentiation," and applies this concept to man as well.

At this point, I should like to say a little more about the distortion or even destruction of the space and time system through domestication. Most free-living wild animals may be said—to put it simply—to move within a system of biologically important fixed points and along definite lines of communication (the so-called " tracks "), keeping to a timetable which they often follow to the minute (cf. Chapter 2). The regularity with which the wild animal moves within this space and time system often puts one in mind of a railway carriage running along rails. Zoo experience stresses a difference here between wild and domestic animals as strongly marked as in the flight behaviour. While the domestic animal often shows hardly any reaction to a change of place, the wild animal's reaction may be catastrophic.

We have already shown how the time factor in the space and time pattern may completely disappear through domestication, in the reference to the loss of seasonal breeding. The spatial emancipation of the domestic animal is often as complete as its emancipation from time. In the zoo where re-arrangements of wild and domestic animals are of daily occurrence, the difference in this respect is very striking.

Just as the wild animal differentiates and demarcates its individual living area, the territory, in various ways, it also has a tendency to alter its living space subjectively in captivity. In either place the wild animal lives so to speak in portion of space permeated with its individual atmosphere. The European bison for instance has its own particular demarcation trees, to which it applies its scent marks according to an invariable ritual. It has no specialized scent organs, in contrast to many other Bovidae, especially antelopes, whose secretion is used for marking, so it uses its own urine as its chief scent, in a similar manner to the brown bear. To begin with, a trough-like depression in the ground, a urine bath, is made at a place in the territory chosen in advance, in which the bull responsible for the marking wallows vigorously, not without difficulty because of its hump, supported by the elongated spinal processes. Yet this hump, with its curve, characteristic of the European bison, has got to be strongly impregnated with scent. To do so, this powerful animal, weighing nearly a ton, twists and turns painfully from one side to the other. When, after obvious exertion, it succeeds, the bison stands up—front legs first, in contrast to the domestic cattle. Then it solemnly walks over to its boundary tree, and with an impressive display of strength, rubs its well-impregnated hump at the appointed place on the chosen tree trunk, which thus acquires a maximum amount of individual scent.

The horns are also carefully rubbed against the same trunk, both the front and back surfaces. In captivity, when marking behaviour is often hypertrophied in all animals, this is sometimes done so vigorously that the horns are thereby abnormally worn down.

What a difference, compared with domestic, especially our European cattle which for countless generations have changed their grounds for reasons of purely human convenience, and not from reasons of biological necessity, such as over-population. Changing domestic animals' quarters in the zoo is child's play compared with moving full-grown wild animals, which often do not survive a shock of this kind.

In 1950 we had an experience of this sort, to quote just one typical instance, with an African elephant, resident for 19 years in Basle Zoo. It had to be transferred to a foreign zoo for urgent reasons. The tame elephant walked with comparatively little trouble into the box wagon of suitable dimensions, but no sooner was it inside, than it started to get violent. This became so bad that it caused so-called lumbago, a myoglobinuria, to which unfortunately the valuable animal succumbed immediately after arriving at its destination.

Among those who capture wild animals, it is well known that among many of the bigger species, e.g. moose, some antelopes, giraffes, etc. it would be senseless and irresponsible to capture full-grown specimens since, from experience, adults do not survive transplanting from their accustomed habitat to a strange one.

There are no such difficulties with domestic animals, however. The more domesticated they are, the more pronounced is their spatial emancipation. The more primitive (the less domesticated), the more noticeable are the rudiments of its spatial dependence. Camels, for example, relatively undomesticated animals, even in the zoo react more obviously to change of quarters than thoroughbred racehorses that are always being transported from one racecourse to another, or than stud bulls and cows that are taken to markets and shows.

I found an extreme development of this spatial emancipation among chickens kept in Central Africa, and carried on safari as living provisions. On the march, the chickens are carried on the heads of the black porters in light baskets. When camp is set up at midday the hens are allowed out immediately on arrival, and find their own food—termites, seeds and the like. They never stray far from camp, which becomes the daily focus of their new habitat. At night they are driven into crates and when wanted, are snared.

Even here there are exceptions: in the case of the carrier pigeon, the original spatial attachment was not only preserved, but even increased by breeding, since its value as a domestic creature depends on its extreme faithfulness to the pigeon loft. Otherwise most domestic animals are emancipated from space, whereas wild animals are fixed to it.

Changing the quarters of a roe deer—a wild animal—is hardly to be compared with changing a sheep's or a goat's. A roe deer, which has been reared on a farm, and allowed complete freedom, may sometimes turn into a ferocious creature if it is taken to a zoo and put into an enclosure, a habitat foreign to it. The result of this may be serious injury, fractures, *etc.* The strange surroundings have a harmful effect and cause almost panic reaction.

The wild animal can only get used to new surroundings gradually, and this fundamental fact must be carefully taken into account in the running of zoos. It is for instance a basic mistake to turn a newly arrived animal out of its travelling cage by force as was formerly the rule in order to get it into its new quarters which to it are strange and sinister. It must be given plenty of time to leave the transport box that it has got used to, and which has become its home as it were, and to take its own time, slowly, step by step, before it becomes familiar with its new surroundings, *i.e.* to create its secondary milieu. In the zoo we often find that a newly arrived animal is extremely reluctant to leave its apparently cramped transport quarters.

We can see traces of this spatial attachment so highly significant in the wild animal, among domestic animals, as in the partiality of the horse or cow for its usual stall in the stable or cow shed. The joke about the milk-cart horse enlisted into the cavalry that used to pull up stubbornly outside the front door of its former customers, to the embarrassment of its rider, is based upon facts of this sort.

The domestic animal follows its master all over the globe, and has become thoroughly cosmopolitan. Incidentally I myself think that there are two types of psychically constituted human beings, the cosmopolitan and the home-lover.

To sum up, the domestic animal, compared with the wild animal, may be characterized as a creature that is not bound by specific ceremonial, and is in particular independent of its environment and of the impulse to avoid enemies. This state of affairs creates a situation of morphological, physiological and psychological degeneration, of distortion that may be grotesque, unbiological, even pathological.

As to the consequences which may result for human beings from this state of affairs, there is no time to go into them, although there is no lack of parallel between certain phenomena of civilization and domestication.

9

ANIMAL PSYCHOLOGY IN THE CIRCUS

IT sometimes happens at election times that we at the zoo are telephoned a few days before the elections by someone with a brain-wave who would like to borrow a camel to parade through the streets displaying large placards with : " Vote for So-and-so." Quite apart from the fact that I consider the use of animals for such a purpose undignified, as well as tactless and lacking in good taste, I am generally not in the position to satisfy such wishes.

Often our camels have not left their enclosures, or the zoo, for years. If, by way of exception, *e.g.* because of building operations or imminent births, they have to be transferred to a different enclosure, this is only possible in face of resistance, and under all sorts of difficulties. To parade them round the busy streets of the town would be a dangerous undertaking, as camels would get so excited that it would lead to desperate attempts to escape, stubbornly lying down, diarrhoea, frightful roaring and perhaps even panic.

Hints of this sort are usually treated very sceptically by the people concerned, and my refusal is often considered clear evidence of badly hidden political bias, since circus camels are often paraded through the streets, as well as elephants and many other animals. In this, zoo and circus animals are often basically different. Most zoo animals live year in and year out in familiar surroundings—the cage—in which they may even have been born. Everything outside this small area is strange and sinister. They feel comfortable and safe only in their territory. If they are forcibly removed, disturbances are unavoidable, and in the zoo it is a primary rule to avoid all disturbances.

In the circus it is impossible to avoid disturbances. The circus is always on the move, the animals are unloaded, loaded, trained, accustomed to music, bright lights, mechanical and human noises. For the circus animal, continual change of surroundings is normal, constant. For the zoo animal on the other hand regularity, uniformity and sedentariness are essential.

Naturally there are all kinds of gradations. In many zoos for instance, camels, llamas and elephants are employed as riding animals and are accustomed to walking about outside their cages, even so they are nothing like so emancipated as their colleagues in the circus.

Another essential difference between zoos and circuses lies in the purely quantitative relationship between animals and staff. Zoos cannot normally afford anything like the amount of money that circuses spend on staff. In circuses nearly every animal is almost continually in close contact with people ; the trainer and his assistants, the riding master and grooms, night watchmen and the like. The circus animal is, as it were, never alone. In the zoo on the contrary, each keeper is responsible for a large area containing many animals and he has little or no time to devote to separate individuals,

117

consequently the intimacy of the circus between man and animal is seldom possible. That is why far more liberties can be taken with circus animals than with zoo animals though there are borderline cases. In many zoos there is a kind of circus business as a side-line, so-called zoo circuses in which certain animals perform in the ring as in the circus. The extreme example of this is the St. Louis Zoo (see Chapter 10) in Missouri, U.S.A.

Performances of this kind by trained animals in zoos may serve one of two purposes—the box-office, or the well-being of the performers—according to whether financial or biological considerations are most important. We must distinguish between more or less sensational showmanship, and simple training exercises in the interest of the animals as a branch of occupational therapy. Here there are two kinds of trainers ; artists, and unscrupulous adventurers and money makers.

When in the past I included a circus ring in reconstruction plans for Basle Zoo, the overriding considerations were biological ones. Many animals turn stupid when shut up in cages and left to themselves. Healthy activity of the occupational therapy sort can be a real benefit to them. If by this means larger audiences are attracted, then that is a doubly gratifying accompaniment. This leads to increase in the number of visitors, and to a continuous and healthy stimulus to the trainer, for trainers, like peda-gogues, are only human. In the ring, official visiting days are of daily occurrence, not as in schools, where visiting days are more or less prepared in advance and do not give a true impression of the average school day. There can be no secrets in the ring, nothing for which the trainer and the zoo director cannot account in public.

Here we must mention, however, that the public often misjudges the performance of good animal trainers because it is misled by old fashioned ideas about animals having to be repeatedly subdued, and overlooks two important matters. First, that it is necessary to provide a proper and satisfactory life for the animal kept in artificial surroundings, and secondly, the way animals behave towards each other. In dealing with a larger animal—often even with a dog—it is at times biologically necessary to give it a slap and even a hearty smack at the proper time and with the right intensity, otherwise the animal would consider the man a weakling and take all sorts of liberties with him. This the animal is forced to do on the basis of its natural social rank.

This may perhaps sound brutal to the outsider, but it is based on biological grounds. From the scientific base of training dogs or wild animals in the circus, this necessity is clearly evident. It is a pity that this point of view is disregarded, sometimes even misused, by greenhorns and unscrupulous people. For a long time, unfortunately, scientific interest in the circus was not considered the thing ; unjustly, and to the detriment of the animals. I have published in *Naturwissenschaften* (1938) my experiences in this matter, as a result of many years with trainers and wild animals in circuses, and have since been able to confirm them in all essentials.

The basic importance of the circus for the study of animal psychology is seen in two respects :—

1. *Material*—The circus affords us first and foremost wild animals whose behaviour is of greater value for analysis by the animal psychologist than that

of domestic animals. This material consists of the most conspicuous larger animals (elephants, tigers, lions, bears, sea lions, hyenas, *etc.*), which up till now have been wrongly neglected by animal psychologists (1935).

2. *Treatment of material*—A complete picture of the psychological pattern of a species of animal can only be obtained by eliciting from this animal a maximum of reactions which we can examine under the most varied conditions, and at all conceivable levels of its relations with man, *e.g.* wildness, familiarity, tameness, degree of training and so forth. In the circus we very often have an opportunity of studying uninterruptedly in one individual the whole range of behaviour changes that occur within the extreme limits of the animal's relations with man—freedom (wildness) and the state of being fully trained. This permits of a complete investigation of the origins of the highest and most intimate of the animal–man relationships ; that of being fully trained and the elements of this.

The achievement of training a wild animal in a circus is by no means of subsidiary importance to the animal psychologist, tied to his laboratory ; it is fundamental, since the handling of animals in the circus is a direct opposite to that of the usual experiments in animal psychology. Kuckuck (1936) is absolutely right in saying: " Circus training is the exact opposite of an animal psychology experiment." In other words, in the circus facts can be observed that must be quite foreign to the experimental animal psychologist in the laboratory. For this reason the laboratory by itself can never be expected to give us anything other than a very incomplete and one-sided picture of the animal's mind.

It is a common idea that in circuses, the animals are only trained to perform those acts—of no interest to scientists—which have show value for the public, while in the animal psychology laboratory, the animal's capacities are put to scientifically useful purposes. Obvious though this judgement of the importance of an animal's performance for the circus or the laboratory may seem, in both cases it is wide of the mark as far as the really important differences are concerned.

Both the animal psychology laboratory experiment and the circus performance seem to have in common one particular method of influencing animal behaviour ; training. To be sure, the experimenter often thinks he is not training his animals at all. In reality, all the usual experiments presuppose some training (maze, puzzle box, alternative choice apparatus, experiments using threads or obstacles, counting tests, *etc.*). Since, however, the process of training often takes place during so-called preliminary tests, they imagine, sometimes quite erroneously, that the actual experiments have nothing to do with training.

Four different scientific fields have been constantly drawn upon for training ; namely, *animal psychology* as for example in the experiments just mentioned (*cf.* W. Fischel, 1938) ; *brain physiology* (for studying deficiency phenomena after excision, for the purpose of localization, *cf.* O. Kalischer, 1907 ; Glees and Cole, 1950) ; *physiology of the sense organs* (*e.g.* the study of bees by von Frisch, 1948 ; or of minnows by Dijkgraaf and Verheijen, 1950) ; and *genetics* (for the study of the problem of the inheritance of acquired characteristics, to which training activities certainly belong ; for example McDougall's rat experiments, 1930). Whenever scientific investigation concerns itself with training, it looks upon it merely as a side-line, and not as a phenomenon of interest in itself. " Little attention was paid to the course of training itself, only the result was important," as G. Hafen (1935) says with truth in her work, in which the first attempt was made to understand the psychology of laboratory training.

Training, however often it is used as a scientific auxiliary, was not a laboratory invention, but represents an age-old way of managing animals ; such as men at a primitive stage of civilization and close to nature used, as well as trainers in the modern circus still intuitively use today. Training as a scientific aid is not yet a hundred years old ; as far as the fundamentals are concerned, these have certainly been borrowed from the circus or its fore-runners. Since then there has been considerable confusion over the concept of training. Training a circus animal means teaching it, by specialized handling and continuous use of effective cues, until it performs certain actions at a special personal signal. These actions, in their elements, are well known to the animal, but in freedom they would never be caused by the same stimuli, and never performed under the same conditions. By scientific training, on the contrary, is meant the foundation of associations (usually impersonal), the formation of conditioned reflexes and the creation of artificial connections.

It is obvious that here are two fundamentally different things, *e.g.* getting a sea-lion to juggle with a burning torch, or a bee to fly to a specially coloured feeding trough. It would be inappropriate to call two such contrasting phenomena simply training, as this would entail a widening of the concept of training, which would lead to misunderstanding. Indeed, French authorities deliberately distinguish between " apprentissage " (scientific training) and " dressage " (circus training). Knoll (1919) found out that it would be going too far to speak of training if, for instance, it was only a question of an artificial association with a special colour and an insect. The term training, however, has come to stay. Thus it is necessary to distinguish between two extreme forms of training in its wider sense. They may be characterized diagrammatically as follows :

| Training in the wider sense | Scientific training (apprentissage) | Without personal signals to perform the required action | Without the use of emotions. Maximum elimination of the animal–man relationship |
| | Circus training (dressage) | With personal signals to perform the required action | With the use of emotions. Maximum intensification of the animal–man relationship |

Our interest in this connection lies in the long-neglected so-called circus training, characterized by moments of intense emotion, that is by a feature which is excluded so far as possible in the scientific training experiment. By the term training we understand a current process not yet concluded.

Training is not a final state, but a transition.   Not until training has followed a successful course does the animal reach the relatively final and higher, or even highest, level of behaviour towards man ;   the state of completed training.   This state is comparatively stable, yet even here there is a definite tendency for the effects of training—the sum of which constitutes the fully trained state—to wear off gradually, unless care is taken to repeat the actions taught during training at short intervals.

I t is a universal rule that animal behaviour shows a marked tendency to fall from the level of being strongly influenced by man to a level where the influence is weaker.   The trained animal gradually loses the powers it has been taught to acquire by man ;   the tame animal turns wild again ;   the captured animal tries to escape, at least at first ;   the animal living in the wild tries to forestall man's approach to it.   The animal has a tendency not only to avoid the animal–man relationship (flight tendency), but to reduce the artificially contrived relationships built up by man.

The process of circus training is not a simple one, but is a complex of several partly overlapping individual processes.   Broadly speaking, three main divisions may easily be distinguished :—

1. The aim that the desired exercise in training should be understood.
2. The overcoming of inhibitions and resistance.
3. The aim of physical adaptation.

Let us first of all consider the overcoming of inhibitions and resistance, which are of the greatest importance for a correct estimate of an animal's capabilities.   From the fact that an animal, even in experiments, fails to carry out a required exercise, it should not be assumed that it has not understood this task.   Very often the performance of an action already understood is rendered impossible through various inhibitions and resistances.   Understanding therefore is not everything.

Among the inhibitions under discussion we must include a type of passive inability.   Inhibitions comprised in this are those conditioned by flight or milieu.   A lion for instance may long have understood that it ought to go to a particular point in the ring.   It cannot do so if a man is standing near it, or if in doing so it would have to pass by its trainer at a distance of less than its flight distance (*cf*. below).

Inhibitions conditioned by milieu are connected with the fact that strange or unaccustomed surroundings are often in themselves enough to induce serious feelings of insecurity or anxiety in an animal.   During encounters with members of its own or other species, an animal that is in familiar surroundings acts more freely and successfully than in unfamiliar ones.   As in the case of inhibition caused by flight nothing can be done by strength at all : the animal must be calmed and will gradually familiarize itself with the peculiarities of its strange surroundings.

It is quite another matter with what we may call resistance.   Here it is not so much a question of inability, as unwillingness—active opposition, without biological necessity.   This sometimes requires the use of suitable punishment, with frequently surprising results in such situations.   As we shall see later on, the trainer must impose his will unequivocally.   The animal would consider his indulgence as weakness, and among the larger animals any manifestation of weakness may have serious consequences, owing to their continual assessment of social rank.

The third facultative phase of the training process, aiming at physical adaptation, is necessary in those cases in which the animal is not physically able to perform the requisite actions. In this case, *e.g.* certain performances of sea-lions and elephants, physical training is required. Only after this has been successfully accomplished can training be concluded, and the fully trained state be achieved.

The three phases just discussed do not in practice succeed each other in this order ; in fact, they more or less overlap. Nevertheless in many cases a limit may be fixed, permitting us to distinguish within the whole training period between a primary period (from the start of training to the attainment of understanding) ; a secondary (attainment of understanding to the removal of inhibitions and opposition), and a tertiary (until adaptation is achieved). To characterize the progress in training single individuals and species we obtain phases which may be directly compared within some limits. These are analogous to the periods of settling down and taming previously discussed (Hediger, 1935b).

Consideration of the first phase of training, up to the achievement of the animal's understanding of the training exercises presented to it, leads to several facts of a fundamental nature which could never have been arrived at by the usual handling of animals in the laboratory. In this first phase the question is to make the animal understand when and how it should perform the required actions. Since there is no possibility of direct understanding between man and animal through speech—as there is between two men—one must find some other way of doing it, and that is by teaching it the meaning of certain signals which have emotional value. It is true there are also some signals which the animal understands properly from the start, and are, so to speak, unspecific as for instance single sounds and gestures.

The study of these primary, comprehensible human signals (expression phenomena), and of their effects, through which their own basic significance is made clear, is a subject in its own right which cannot be dealt with any further here. The primary comprehensible signals are not enough in any case to make clear to the animals all the necessary details for training. It must in addition learn the meaning of a large number of signals, the meaning of which can only be grasped secondarily, as a result of individual experience. Therefore there is no alternative except to force it to perform actions which it will later carry out as a result of signals alone. Only in this way can it be made to understand what is required of it. At the same time it must be treated as gently and sympathetically as possible, since excitement is harmful to a successful course of training.

This method of training is called " putting it through the action " ; forcing the animal to make passive movements which O. Koehler (1928, p. 921) rightly describes as being the explanation of the success of most circus tricks. We may distinguish between various forms of this method, which are applied according to the species of animal, or to the trainer. The purest form of " putting through " is used in training elephants. Here the " forcing of passive movements " is seen at its most striking, since human strength is often so insufficient that pulleys have to be used, or the help of already trained elephants must be sought. If an untrained elephant has to learn to " sit down ", there is no alternative except to force it down gently on to a barrel that serves as a stool and then raise up its front legs ; the animal must not be excited. If a brown bear has to learn to walk on its fore paws, it must allow itself to be held up by its hind legs for a time.

This pure form of the " putting through " method demands direct handling. This again is only possible with completely tame animals, *i.e.* with animals whose flight distance equals nought ; whose tendency to escape from man has been overcome, and which may thus be touched without risk. With species of animals that for biological reasons respond readily to food stimuli (as they eat little but often in natural conditions), particularly sea-lions (*cf.* Spindler and Bluhm, 1934), and bears (*cf.* Kuckuck, 1936) much may often be accomplished without the necessity for direct handling ; they can be enticed to any desired spot—pedestals, pyramids and the like—by holding out food. Thanks to the irresistible attraction of a food bait, a kind of passive movement can be elicited by this means.

That form of the " putting through " method, although not at first sight recognizable as such, in which the animal is neither directly touched, nor tempted with food, is most interesting psychologically. This method is most frequently used with beasts of prey such as lions and tigers, that are not tame.

It is biologically significant that, in contrast to bears, they feed at long intervals, but then in great quantities : food stimuli clearly have far less importance in their case. Generally, taming of some sort precedes training: many trainers, however, prefer to start training the big cats when they still show considerable flight and critical reactions (Hediger, 1934).

According to definition, flight reaction occurs in every wild animal, or in an animal adapted to captivity, whenever man approaches to within the characteristic flight distance of that animal (*cf.* Chapter 4). With animals born in captivity, or accustomed to it, the flight distance is considerably less than under normal conditions in freedom. Strictly speaking, one should therefore speak of secondary flight distance in the circus, to distinguish it from primary flight distance in freedom. In any case, the element of wildness (flight-tendency) present in the animal at the beginning of its training puts the trainer in the position of being able to drive the animal away from him to any point in the ring he desires, *i.e.* to set off its flight reaction, simply by overstepping the significant flight distance of the animal, or allowing his whip to do so. The whip obviously signifies to the animal the extension of a human extremity ; the personality of the trainer is likewise projected into the whip, and its movements are part of his gestures.

In addition to this fundamental possibility of driving the new wild animal away to any desired point by releasing its flight reaction, the tamer has another basic possibility, of drawing the animal towards him from any point by releasing its so-called critical reaction. This occurs whenever the man approaches a wild animal (or one accustomed to captivity), which is prevented from escape, to a distance less than its characteristic " critical distance." In the case of the big cats, this distance plays an exceedingly important part, is fixed, and may be determined to within centimetres. The critical reaction consists of a change from flight to attack, never with the character of an active offensive, but always of a defensive, emergency nature. With the aid of his auxiliary extremities (whip, stick) the trainer in the circus ring is easily able to drive an animal into a corner, *i.e.* to cut off its avenue of escape, and then to release the critical reaction by approaching too near (to less than the critical distance away).

In contrast to flight reaction, which may occur in any direction, radially, critical reaction has the advantage of being aimed in one direction. As an attack (*i.e.* emergency defence) it is always aimed straight at the trainer. This permits him to place obstacles, in the form of some kind of training apparatus (*e.g.* pedestals), between himself and the attacking animal. Thus the animal has no alternative except to make its way over the training apparatus under the influence of critical reaction. It is true that an animal can, by releasing its flight reaction, be driven over a barrier set up against the bars and which stands in its path : but whenever it is desired that the animal should move in a definite direction, it is better to use its critical reaction.

When for instance a lion that is attacking, under the influence of critical reaction, mounts a pedestal that blocks his way to the trainer, the latter is then in the position of being able to check the attack (or critical reaction) any time he wishes, even when the lion is standing on the pedestal. In order to interrupt the critical reaction, and stop the attack, the trainer has only to step back a little more than the critical distance of the approaching animal, thus removing the motive for attack from it. The result of this manoeuvre is that the lion then remains on the pedestal and there calms down until the trainer drives it off again by a further approach that releases its flight reaction when escape is possible. The pedestal may thus acquire the character of a resting place for the animal, through repetition.

With reference to the types of subjective world described by Brock (1934) as home, shelter and refuge, we may clearly recognize in the circus animal, *e.g.* lion, three types of home—of the first order (= cage it lives in) : of the second (= individual " place " in the ring) : of the third (= pedestal ; pyramid structure)—as places of relative security which attract the animal. The strongest attraction comes from the first order home ; on their way back from the ring to their living quarters the animals often need little driving, in opposition to the way to the ring, where they then often go to their " place " or substitute home of their own accord. From there to the least attractive third order home they must often be driven again.

Instead of a pedestal an animal can equally well be forced on to a see-saw, or a pyramid, and back to its place again without being touched, simply by a skilful use of its flight and critical reactions. It often looks as if the trainer were pulling or pushing the animal to the desired spot by invisible wires.

Displays of this kind, based on fundamental biological knowledge, are in sharp contrast to such showy performances as bears riding motor-cycles, elephants sword-fighting, or animals firing guns, to quote only a few particularly grotesque examples.

All this discussion about releasing flight or critical reactions may seem much too cut and dried, much too simple—yet the performances are matters of course to any good animal trainer ; he simply does not give a special name to the different distances and reactions that seem so obvious to him, and uses them spontaneously ; but may not a musician perform marvels without bothering about the frequency of vibration of the sounds he produces ? Just as a musical performance can be analysed physically, *i.e.* accoustically, a trained animal performance can be analysed biologically. The trainer works intuitively, like an artist, spontaneously using all sorts of biological distances and reactions as the result of his experience ; the animal psychologist on the other hand measures, names and describes them.

It requires extraordinary attention and concentration, and is a feat of great difficulty to work, not with single animals, but with a whole group, as is usually the case. The work of the Swiss trainer, Vojtech Trubka, is outstanding, particularly in the field of biological training of beasts of prey. I have made a special investigation of this artist's work, during a fairly extensive study of circuses, and will return to it later.

If we consider the various forms of the " putting through " method, of the imposition of passive movements, from vigorous handling of elephants to long-distance control without contact, of movements under biological coercion of lions or tigers, through provocation of flight and critical reaction, we can in spite of all dissimilarity still find one extremely important feature in common and that we must now treat at greater length.

All the so-called aids (coercive interference), such as the raising of a leg by direct handling or regular provocation of an attack by release of critical reaction, are in the course of training (*i.e.* of the repetition of these aids) gradually cut down, but with the preservation of the original emotional content. After several repetitions, these aids still appear effective, even when they are only hints or symbols of their previous form. Instead of laboriously raising a leg, a slight pat is enough ; instead of a literal violation of the flight distance, a significant movement of hand or whip in the direction of the animal suffices. Eventually the pat becomes a faint gesture, the whip movement a provocative look. In other words, the original essential means of coercion turns into a symbolic signal. Within these signals, always closely linked to the personality and expression of the trainer, are concealed, in the last analysis, the significance and effects of the earlier forms preceding these signals, and so back to the original coercion. Finally, these personal signals often pass unnoticed by the human observer in the general expressions of the trainer, blending with them so that they no longer appear at all obvious.

In respect of its ability to interpret expression and the training signals connected with emotional stimuli, the animal is often far superior to man, at least in so far as it can distinguish between true and false straight off. In the majority of cases therefore, human play-acting and make-believe misfire with the animal during training. In order to obtain a satisfactory performance, the appropriate expression and the training signals directly connected with it must be genuine ; these signals must really relate to the emotional content which the animals originally had. As a rule, the animal will not respond to empty gestures and shallow mimicry.

The trainer must therefore be prepared to concentrate inwardly on the required actions. It must not be assumed that in a circus performance with wild animals, the trainer is standing passively in the background, and merely letting the animals do the work, just because so little of his co-operation can be noticed. Without the inner collaboration and concentration of the trainer, which can usually be realized only at close quarters, but which often becomes conspicuous through typical accompanying gestures, nothing would be possible. This is the reason for the enormous strain attendant on each good performance. Any lack of concentration, indisposition, or depression on the part of the trainer is at once reflected in the work done by his animals. During a large number of performances, which I often recorded by word and picture in long continuous series, a repeated cause for bad work by the animals was that the trainer was suffering from dental pain, or was otherwise indisposed. Naturally the performances are also largely dependent on the actual mood of the animal.

Before we go more closely into the theoretical significance of the signals described, let us look at some of the most important complications, in the cases of lions and tigers, that usually arise in a group of predatory animals in the course of time. The behaviour of animals in a training group may be subject to extraordinary variations within the specific boundaries of each species, and will never be understood, unless at least three characteristic stages are distinguished, conditioned by the age of the animal.

It may happen that an animal in the same situation and under the same stimulus, reacts in one way at one time, and at another (*i.e.* at a different stage of life) reacts in exactly the opposite way. I saw a case where a trainer was unable to walk round the pyramid, which occupied nearly the whole of the diameter of the ring, because the animal sitting closest to the iron bars ran away whenever the trainer tried to pass between it and the railings. In another group (with animals at a more advanced stage of life) the trainer was prevented from walking round the pyramid because the animal occupying a similar position kept on attacking him. This apparent contradiction and irregularity of behaviour can nevertheless be understood if one takes into account the actual stage of life at the time. These phenomena, of course, must not be confused with individual differences of behaviour.

Lions and tigers—to keep to our examples—are usually started on their training at the age of from two to four years (and for preference caught wild, not born in captivity) ; that means generally before puberty. This first stage, as we have seen, is very often marked by a definite tendency to escape from man. When it is not present it can be artificially provoked by suitable irritant treatment. A group of young animals like these is not yet materially organized socially ; it represents from the social aspect a sum of almost equal ranking individuals, each of which is in character already a single personality.

During their intercourse with men these animals usually lose their flight tendency gradually, but completely, under proper treatment ; thus they grow tame. The emotional link, the attachment between animal and man, is then at its strongest. Tactile stimuli play a large part ; the animals love to be scratched and stroked, and the trainer can move freely among them and do practically anything he likes with them since flight and critical distance have disappeared in the tame animal. Such animals are generally young adults, are approaching puberty, or have recently emerged from it ; they belong to the second stage.

With the third stage, fully adult animals typically show a definite, often highly pronounced tendency to keep at a distance again, which however has nothing to do with flight tendency. The attachment often disappears and is replaced by a particular kind of animosity or inaccessibility. This is caused by the great social transformations that have meanwhile been taking place between animals in the group; an extremely strict and precise social organization (ranking) has been set up. Every animal now stands in a definite, often highly complex, relationship with every other animal in the group. One particularly prominent animal has fought for itself the position of leader, so to speak, the highest rank inside the group (α-position). A second animal has managed to fight its way to the next highest place (β-position). The lowest place is taken by the animal of lowest social rank and must always give way before all the others (ω-position). Moreover, such animals of very low social standing often show a striking readiness to oblige, and a submissiveness to the trainer, which considerably eases his work with them, so that these animals themselves can be made most use of, and for the hardest tasks. On the other hand, individuals of high social rank are less amenable, and thus more difficult to work with. Each animal at this stage of life must always be on the look-out to improve the social position it has fought to gain at every opportunity, since this is of the greatest importance for the kind of life it will lead. In any case, it must always be ready to defend its position. There are periods when social disputes of this kind are more frequent and violent, but they are interspersed with periods of social calm.

These social relationships, that were quite unknown to the young animal, inevitably lead to an almost perpetual state of readiness for fight, or at least to permanent unapproachability. The socially organized animal must literally not be approached too closely, for it would immediately take it as an attack on its rank. The older an animal, the more sensitive it usually is on this point. The area immediately around the animal is a kind of no man's land. Usually only individuals linked by special friendly relations, among whom may sometimes be included the trainer, are permitted to approach each other without a fight.

A number of consequences, of the utmost importance to the trainer, emerge from the social relationships just mentioned. If to animals of the first stage he has the significance of an enemy of a different species, with flight reaction occurring at his approach, and if to second stage animals, the significance of an older or social companion (Lorenz, 1935), the trainer certainly appears to animals of the third stage, at least in the majority of cases, as a social rival. Hence the only course open to him is to assume this part with success, and to continue to play it skilfully. The most important requisite for this is that, at all costs, the trainer must ensure that he takes the position of undisputed leader, and maintains it resolutely. A trainer among a group of predatory animals is unthinkable except as the chief individual, the α-animal. Only by playing his part properly, with the greatest concentration, and with complete consistency, can he succeed with his animals. The slightest sign of weakness, of fear, or anxiety, a stumble, and so on, might be the signal for an attack by the animals, his social rivals, since they are always aspiring to take his place in the group.

Thus, during disputes about social position among animals of the same species, it is most useful for the trainer that it is less a question of physical superiority, which would lead to a bitter fight to the death, than of assured and imperturbable appearance, in fact, of imposing manner. When two animals meet, the one that can intimidate the other will be accepted as socially superior (as Schjelderup showed in his classic research on hens). In this way the trainer can often succeed in compelling the animals to recognize his social superiority by energetic behaviour, rather than by dangerous fighting. It demands incredible discipline, if he is to assert at all costs his superior position by vigorous reprimands when, for instance, fighting breaks out among the animals, and never to act out of character however ticklish the situation may be.

Certain considerations of space are important in the course and outcome of socially conditioned fights and quarrels. Disputes of this kind are greatly affected by the type of ground : whether it is familiar, or strange ; their own, or already occupied. The trainer, working with adult predatory animals, does not enter the cage before them unintentionally ; he is, so to speak, demonstratively taking possession. The animals, following him in, find the territory already occupied. Similar problems of space must be watched for closely, when mixed groups are put together. When bears and lions, for example, are being trained together, the more nervous bears are let into the cage ahead of the more aggressive lions.

From all this, it is easy to see that the older his animals are the more difficult is the trainer's task. The public believe, mistakenly, that the older animals are quite harmless. Every moment that the trainer spends during the performance in the cage occupied by trained adult predatory animals, demands the accomplishment of a fourfold task. 1. The safety of his own person. 2. The safety of the animals, *i.e.* the stifling of their ever-smouldering, socially-conditioned pugnacity. 3. Inciting the animals to perform the required actions. 4. Artistic effect upon the audience.

The safety of his own person and of the animals again depends on unremitting attention : 1. To specific characteristics (*e.g.* of tigers or lions) ; 2. To individual characteristics (character) ; 3. To social characteristics (friendly or unfriendly relations among the animals, social rank) ; 4. To the prevailing moods ; 5. To the actual situation of the moment. Here it is impossible to illustrate with examples these theoretical data, drawn partly from exhaustive observations and partly from personal experiences. In any case, the performance of a good group of adult animals makes extraordinary claims upon their trainer. It seems incredible that he can keep his eye on so much. Each step, almost each single movement, is made deliberately, has a meaning, and forms part of a most complicated and precise interplay of distance, movement and counter movement—never without danger.

After this reference to some of the peculiarities of training and performance of trained animals, let us return to the point that basically all circus performances with wild animals depend on special kinds of affective signals. During the course of training the animal learns a lot of human signals and expressions and, conversely, the trainer develops an understanding of the animal's expressions and signals. Working with his animals teaches the good trainer to interpret the finer shades of the animal's expression swiftly and surely, and a considerable part of the trainer's art lies in its masterly interpretation. All that interests us at the moment is the signals given by the man, and their effect on the animal.

Figure 14.—Chimpanzee ' Pancho ' at the St. Louis Zoo, Missouri, driving his electric jeep, puts it into reverse after a head-on collision with the ramp.

Figure 15.—Tame porpoise (Tursiops truncatus) in the Marine Studios at Marineland, Florida, jumping out of the water at feeding time and taking the fish out of the keeper's hand. This unique establishment offers an inexhaustible wealth of new fields for observation to the animal psychologist.

Figure 16a.—*Beginning of a rival* fighting *between two rattlesnakes* (Crotalus ruber), *as observed in San Diego Zoo by Charles E. Shaw and put on permanent record by the Zoo photographer G. E. Kirkpatrick in December, 1947, in these unique documentary pictures.*

Figure 16b.—*The two fighting males, whose strange movements have hitherto been completely misinterpreted by most observers as mating preliminaries, are both looking in the same direction. Their heads form an angle of approximately 45° to the rest of their bodies.*

It is obvious that different species of animals show very different reactions to training, and even the general features of the most important species cannot be dealt with here. Brief mention can only be made here of two extreme cases. Most wild circus animals do not understand immediately what they are required to perform, it dawns on them gradually, although occasionally a sudden decisive burst occurs. Let us take the lion as an example. During jumping over the bar, reached at the beginning by the release of flight reaction, the obstacle must always be placed radially to the cage bars, so as to prevent the lion from sneaking off sideways, necessitating its jumping over the barrier, when forced in that direction by the trainer. After continuous repetition, the lion at last begins to understand because its positive achievements are always rewarded with praise. Then the obstacle can be put up on its own, some distance away from the cage bars. The animal has now learnt that the barrier must be jumped over, not avoided. For various reasons, failures frequently occur, especially at the outset, and then the training begins all over again.

In contrast to this dawning understanding, there is also the sudden realization of what is required at training. This phenomenon is very pronounced in sea-lions, as Spindler and Bluhm (1934) confirm in their informative investigation of this species. For this reason, sea-lions can be taught to perform actions that cannot be learnt gradually, but only at once, or not at all ; for example, certain juggling tricks, for which sea-lions are particularly well adapted for biological reasons.

In this connection, however, the important thing is not the characteristic type of understanding in a single species, but the basic kind common to all circus training. Details have been given of the arguments between trainer and animal, aimed at the desired performance, and basically consisting of play and counter-play, with signals dependent on expression, and based on emotional reaction (of mimetic, gesticulatory, and audible kinds ; even olfactory and other kinds). The trainer, thanks to his experience, recognizes the tiger's challenging look at his partner as the prelude to a fight, and responds to it, generally unconsciously, with a gentle call, a threatening move of the head, or a slight flick of the whip at the pugnacious tiger. This faint expressive movement of the trainer contains all that is needed to nip in the bud the tiger's rising pugnacity, and to keep him sitting quietly in his place. This movement is accepted by the animal as an agreed signal, moreover it contains the effect of a command. Here there is a marked difference between the personal signal, with its emotional implications, and the impersonal mechanical signal, often used in scientific experiments.

By these faint signs, the animals are persuaded, or forced, to single tricks, or complicated trained acts. Finally they react according to the trainer's intentions, sometimes in a whole series of actions. Noisy shouting and whip-cracking or even firing revolvers (blank) are intended very often only for the audience, although sometimes used as a spur to overcome the animals' resistance, which stands in the way of the performing of an action thoroughly understood by the animal. In difficult situations, however, such as fights over social rank, these symbolic signs are no longer sufficient ; animal and man have necessarily to use stronger measures.

In general, it may be said that the understanding necessary on the animal's part for a successful training process, is not as a rule an intellectual understanding of complex causative associations, but rather an emotional understanding of sympathetic expressive movements of the man. It is hard to draw a clear line between expression proper, and those minimal (traditional) signals, subsidiary aids ; the one merges into the other. The subsidiary aids develop into expressive movements, that originally accompanied the chief aids. This does not mean that the intellectual sphere plays no part in all circus training. The indispensable basis, however, for the formation of these important actions in the sphere of emotions is the peculiarity, the intimacy, of the animal–man relationship, so very characteristic of all circus training.

The previously mentioned training of predatory animals, as practised by V. Trubka, is an art in the true sense; but its elements are not colours or shapes or sounds, but the movements and emotions of wild animals. Out of this unique basic material is created the work of art that this sort of performance is, transient, fleeting even, but none the less sublime. In order that so excellent a performance can be appreciated and understood, it appears desirable to keep sight of the basic facts we have just presented. We do not go to the theatre or the concert completely uneducated, but take a certain amount of literary or musical understanding in with us. The better we are prepared beforehand, the greater our enjoyment of the fare provided, and the deeper the experience.

Trubka's former group of six tigers consisted of two hostile sub-groups, the members of which—Sahib, Kora and Rani on the one hand and the gigantic Korfu with Mirza and Jasmi on the other—had to be kept carefully apart, at all times, and everywhere, even in Home I, their living cage, or else had to be watched with the greatest attention, if mortal fights were to be avoided. Thanks to the trainer's previously mentioned mental superiority, he was able to order all the four-legged members of his fellow species about, including the chief tiger Korfu, and Mirza, his nearest rival ; to get them into their places, and then to make them perform their individual tricks. These trained performances, called tricks in circus jargon, included the pyramid, when these magnificent animals first offered their backs to the audience, then their fronts, finally sitting up on their haunches. Then came Jasmi's splendid leap over Korfu and through the hoops ; Kora's jumps through the paper-covered hoops, and Mirza's through the canvas tube.

The method of locomotion so characteristic in many situations of these animals—leaping—appeared during these tricks in an extremely significant way. Mirza's sideways roll, connected directly with natural playing, was outstanding, and so, above all else, was Sahib's walking backwards round the whole ring, during which the difference between human and animal position was strikingly illustrated. Then, too, the four tigers forced to lie down next to each other, followed by Jasmi jumping across them all—not shown at every performance—was a superb act.

All this happens in about ten minutes, but forms only a part of the performance. These are just the highlights in the wonderfully harmonious interplay of animal and man. A considerable part of Trubka's work consists of masterly judgement of the biological distances and symbols (aids) mentioned. His performance is a continuous and very delicate interplay of symbolic provocations and reactions of the animal, an elegant weaving together of movement and counter movement. His mastery of the whip, for instance, is in itself an artistic performance, and the way in which Trubka moves among the paws attacking him on all sides reaches the limits of human powers of calculation. An additional factor in all this is that the master's own movements so fit in with those of his animals that they blend into a perfect harmony with them. The observer can often hardly distinguish between deliberate, carefully calculated movements and involuntary accompanying ones. This is the secret of the special style of this beautiful performance.

Apart from the actual training work in the circus, there may sometimes occur unbelievable actions by animals, based on the close, emotional connection between man and animal, and these are often mistakenly considered as evidence of high intellectual power. These include, for example, those cases in which the trainer is defended by one animal against others, or when an animal is punished or corrected independently by one of its fellows, when it will not perform properly. Among other instances, I noticed one in which a Himalayan black bear was about to try to escape over the cage bars in the ring, in consequence of a sudden fright. In spite of the efforts of his assistants, the trainer was unable to bring the bear down from the topmost bars. Then a European brown bear, which was on unusually intimate terms with the trainer, interfered. After a short pause at the bars (normally climbing up was forbidden) the bear clambered up over the two men, bit the runaway on the leg, hauled it down to the floor of the ring, and only let it go after biting and spanking it. Such an astonishing sort of deliberate interference in a situation can be observed comparatively frequently among circus animals.

Many similar examples of animal behaviour are wrongly attributed to "trainers' tales" or "gamekeepers' romances", etc., simply because they appear of outstanding intelligence, and because these affective reactions have always been observed only by trainers, hunters, amateurs, etc. and never under laboratory conditions.

In the case of animals communicating by knocking (cf. Chapter 11) a phenomenon that once astounded the whole biological world, the interpretation of the human signals by the animal were finally recognized as an almost catastrophic source of errors. Strangely enough, only one of the two possible deductions from this recognition, and always the negative one, was drawn, namely the hindering of the intimate animal–man relationship in the normal animal psychology experiment, and thus the elimination where possible of human cues as the cause for pseudo-capacities.

This painstakingly thorough elimination, which might be called the elimination method, was almost grotesquely over-exaggerated by Pavlov (1926). He deluded himself into thinking that he had banished man with all his upsetting influence from the laboratory and the accurate experiment (Podkopaew, 1926) ; but this elimination is never wholly successful, even with the help of all Pavlov's cunning devices. Every animal psychology experiment is necessarily an encounter between animal and man (cf. also Buytendijk and Plessner's (1935) excellent criticism of Pavlov's methods).

The other positive deduction from the recognition of the sources of error due to the animal–man relationship would be the accurate analysis of these upsetting personal signals, and the explanation of their effect. Seen from this fresh viewpoint, circus training becomes of fundamental importance in this connection. It is clear that these two extreme methods of animal handling show animal behaviour from two fundamentally different aspects. Experimental animal psychology, tied to the laboratory and its usual experimental animals, continually dissects and analyses only one half of the animal's mind, while it often suppresses, and even entirely neglects, the other equally important half, the affective sphere. To that extent, the study of animal behaviour in the circus together with observations in freedom and, under non-experimental conditions, in captivity presents not only a favourable but a truly necessary supplement to the research undertaken in the laboratory.

10

PLAY AND TRAINING

In Chapter 3, I attempted to describe the daily life of the wild animal living in freedom, its two chief occupations were stressed, namely the constant alertness, the everlasting state of being on guard for avoiding enemies, and the implacable necessity for seeking food.

In the zoo, where the animal is protected from surprise attacks by the bars, and where every inmate has ample food provided daily, both these main activities disappear. Thus a vacuum—an occupational blank, as it were—is caused, which may lead in the worst cases to complete boredom. In former days, especially when menagerie animals were kept in isolation in narrow, gloomy dens, there were serious effects as a result of unnatural inactivity. The animal's need for occupation frequently showed itself in a distorted or even morbid manner, resulting in all kinds of stereotyped movements (Hediger, 1950) ; running to and fro hour after hour ; endless walking in circles, or all sorts of silly behaviour and grimacing, even self-mutilation and so on.

These extreme manifestations of boredom are noticeably less frequent in larger cages and enclosures where the animals can be kept as families or groups, yet even today one is sometimes obliged to keep an animal permanently or temporarily on its own in the zoo, either because it cannot get on with its fellow members, or because its mate has died and cannot be replaced. Different species of animals are extraordinarily different as far as the formation of stereotyped behaviour is concerned. Some have great need of activity, others very little. In the higher species, considerable desire for variety is shown; one might say in many cases for entertainment. Where this is not forthcoming, clever animals often make up a game of their own, but the keeper does not always agree with this compensatory solution.

I recently had an example of this in the famous anthropoid ape farm in Florida, founded by Professor Robert Yerkes in 1930, and run by Yale University. In 1951 no less than 67 chimpanzees were living there. More than 50 had been born there, some were of the third generation, and the first representative of the fourth generation was on the way, that is, a baby was expected whose grandparents had been born in this anthropoid ape farm. Many of the chimpanzees live together in larger or smaller groups in roomy communal cages in the open with ample facilities for climbing tall trees. In this area, all kinds of birds, lizards, or even an exciting snake or an amusing insect, are to be seen or even caught. There is no lack at all of amusements here. Other chimpanzees are for some time every day kept busy with the experiments, various tests, selective experiments, *etc*. A chimpanzee that likes nothing better than a sheet of paper and a pencil, may draw a great deal. Hundreds of its sketches are expertly examined by psychologists, many are published.

In addition to these fully occupied chimpanzees there are some, particularly the older ones, which no longer co-operate willingly in experiments, and which are in consequence obviously bored. They have, however, discovered how from time to time that they can enjoy themselves, and so a bad habit has become a tradition in the anthropoid ape farm. These chimpanzees squirt water like a fountain. I would never have imagined that an anthropoid ape could hold such quantities of water in its mouth.

Of course the director, Dr. Lashley, warned me about this at the very start of my visit, and so I did my best to keep out of range of the jets of water during the inspection of this colony of chimpanzees. In the tropical climate of the Florida midsummer, the apes have as much water as they require, and they have developed an almost incredible skill in filling their mouths as quick as lightning, and usually unnoticed, with an unbelievable quantity of water, then, with the most innocent of faces, letting fly without warning at any human being, engrossed in his thoughts, or deep in an argument, that comes their way.

I thought I was proof against such tricks and I managed to make notes, take measurements and photograph without getting any of the dreaded showers. On one occasion I realized that it was time to go to the administrative block to keep an appointment. I had the choice of going directly between a corner cage and a thick clump of bamboo, or take a longer path. Mindful of the warning impressed on me, I took an unobtrusive look at the occupant of the cage, close to which I had already for some time been busy photographing, *etc.* He happened to be an old chimpanzee that was apparently taking no notice of me. He sat there listlessly, with his back to the cage bars, and seemed to be playing with his toes. As a precaution, I looked at his cheeks, which didn't seem to me to be the least bit swollen. So, as if by chance, I passed between his cage and the bamboo grove, keeping as far as possible—some twelve feet—from the cage.

When I had reached the narrowest part, the old chimpanzee suddenly swung round in a flash, reached the railings in one leap, and, unable as I was to retreat any further, drenched me from head to foot in a stream of warm water. It gushed out as if from a hydrant. The cunning ape must have been watching me for some time out of the corners of his eyes, and have taken in his water supply as a precaution; he then squatted hypocritically, acting the innocent so well that he caught me out a hundred per cent. at the spot most favourable to him. Although I did not come out of it with shining colours as an animal psychologist, especially in the presence of so many other chimpanzees, I was nevertheless glad for the old ape's sake, that his joke had come off so well. Who knows how much good this well-prepared incident did to him ? It was no doubt a very welcome break in his life of boredom. For me, it demonstrated once more the necessity of continuous occupation for captive animals, and is another example of an animal inventing a game for its own amusement.

This kind of behaviour clearly shows the animal's great need of occupation, of substitute activity for defence against enemies and the search for food, no longer necessary in the zoo. Naturally, this need is not only found among apes, but in many other animals—sea-lions, for example. In every zoo sea-lions have been seen to take hold of a piece of wood floating in their pond, and playfully tossing it up and catching it again. We can recognize the beginnings of juggling and ball catching in spontaneous play of this kind. According to the keeper, one of the sea-lions had fitted itself up with a regular toy shop in a cleft in the rocks, where it hid all the things it used to play with from time to time. It would spend a lot of time there, and seemed to choose its favourite playthings according to its mood at the time.

Many predators too are very glad of some pleasant interruption of their rather monotonous existence; that is why they are often given wooden balls and the like to play with. Strangely enough, the ungulates' needs are usually underestimated in this respect. In fact, antelopes, wild sheep and goats, as well as Equidae and many other animals, have considerable need of activity, and here much has still to be done in zoos to satisfy it. Birds too, even certain reptiles and fishes, should have something better with which to occupy themselves. Belle J. Benchley (1942, p. 61) records how large macaws play with stones. In her excellent book on anthropoid apes (1942a), she stresses the importance of opportunities for amusement and occupation for these primates.

Many zoo animals that are kept in spacious enclosures in large herds can provide their own specific games at will. They themselves counterbalance the lack of activity caused by captive conditions ; many ungulates for instance do this if the field at their disposal lends itself to it. Some of the most familiar games of this sort are known as " King of the Castle " to behaviour experts. It is one of the most universal animal games, and is incidentally played—with exactly the same rules—by human children. On the way to school, heaps of sand or gravel used in housebuilding or street repairs have an irresistible attraction to children. One of the bolder youngsters climbs to the top and announces : " I'm the King of the Castle " while another equally playful pal tries to pull him down and take possession himself. The same game is enjoyed with equal pleasure by man and animal.

Lonely elephants can get much pleasure out of a plank or a stump of root that is put into their quarters. The young Indian rhinoceros " Gadadhar " in Basle Zoo gets much pleasure every day out of a block of wood, which it tosses and rolls about its cage so vigorously that it has often made its nose bleed, and the block has had to be removed for the time being. Recently, it has been using a solid rubber ball weighing more than 115 lb. which was specially made for it. Games are without any doubt the best and most natural way of filling in the most significant and biologically suitable manner the gap in activity caused by captivity.

We zoo men have certainly no ambitions towards highly artistic performances, but are only concerned with providing for those zoo animals that give us the chance a suitable amount of movement and activity. We want to provide occupational therapy so far as we can for the benefit of our charges, once it has become obvious that long inactivity has a most serious effect, not only on men but on animals.

In Basle Zoo, headkeeper Stemmler has taken immense trouble to entertain the anthropoid apes, the two chimpanzees " Martha " and " Pablo " and the gorilla " Achille ", three times a day at the official feeding times. These inhabitants of the African virgin forest have to sit down properly at table, have their bibs tied on, and are not allowed to start until they get the word. After their meal, they indulge in all sorts of acrobatics, including riding scooters and horizontal bar exercises.

Here play has imperceptibly merged into training practice. In this connection, training simply means disciplined play—and a certain amount of discipline is essential for animals, whose physical strength is greater than man's. This is soon the case with anthropoid apes, as also with the larger predators. Man, with his thin skin, his fragile bones, and his relatively weak strength, simply cannot compete on the same terms with anthropoid apes, tigers or elephants, if he wants to keep out of hospital. A certain amount of discipline is thus unavoidable, but it is in the interests of the animal to give it in the form of simple exercises in training and obedience. For example, one cannot in an emergency suddenly take the temperature of an anthropoid ape that has been left completely to itself ; it would not be used to this contact, and would strenuously resist, as it also would when medicine was offered it. Thus daily civilized sitting down to table for meals, and disciplined games with the keeper, accord priceless advantages, in addition to the beneficial activity of the animal, in the form of keeping an eye on its health and, when necessary, giving it medical treatment.

In some zoos, they go much further than just the necessary training activities, and enter the field of artistic performances and record acts, which are really the province of the circus. Among the most celebrated zoos of this kind is the one at St. Louis, Missouri, U.S.A., which I recently had the opportunity of visiting. This zoo has three circus rings, in which elephants, lions and chimpanzees perform.

Of the 22 chimpanzees in this typical American zoo, 13 have been trained as circus artists and perform three times a day before huge and delighted audiences. Among the so-called stunts, consisting of the most astonishing tricks of training, and nowhere else achieved, are driving motor cycles and jeeps powered by electric motors. A complete car park is at the disposal of these chimpanzees, and at a signal from their trainers, they dash up to the stage and begin a somewhat wild motor race, in the course of which the vehicles may collide.

Traffic jams may occur, for which the only solution is going into reverse, and this these expert drivers do themselves without hesitation. I must however add that I felt some doubt over this performance : are these chimpanzees really such clever drivers that they back on their own, in order to gain a clear passage once more ? Georges Vierheller, the popular manager of the St. Louis Zoo, gave me every opportunity of investigating these apparently fantastic performances. I closely examined these cars and went over to " Pancho ", one of the most gifted chimpanzee drivers. With the greatest self-possession, he took his seat at the wheel, waited for the starting signal, put the car into forward gear in the usual way, and off it went. We had the greatest difficulty in confusing this chimpanzee driver, so as to make him have a collision, or bump into the ramp. He looked round for a moment, then stared at the trainer, who, like a traffic policeman, was waving to him to back. Without watching what he was doing the chimpanzee grasped the lever, put it into reverse, and set off again briskly.

Personally, I first thought this mix-up of creatures from the African virgin forest and modern motor cars somewhat repugnant. When, however, I had seen the whole troupe of chimpanzees waiting, before the curtain rose, for the start of their stage race, I began to see the thing in a different light. These animals are very like fidgety naughty schoolboys. Just before they appear, they play all sorts of pranks on each other to fill up the time. They climb from one car to another, squat on the bonnets and scuffle ; they keep on leaving their cars and chasing each other. They cannot sit still for a moment, but dash around, until you could hardly believe that a proper entrance was possible. On the rise of the curtain, however, each one makes a dash for his " place " and the noisy cheerful procession starts off, ideally suited, so it would seem, to the chimpanzees' irrepressible spirits. Apart from driving cars, these versatile artists take part in many other kinds of activities such as riding ponies as jockeys, doing gymnastics, walking on stilts, dancing, performing acrobatics, *etc.*

Surprising as were these performances to me, going it seemed beyond the extreme limits of the animals' capacity, I came across even more astonishing things in the field of animal training on my journey through the States (Hediger, 1952). The highlight for me was a trained dolphin, a porpoise to be exact (*Tursiops truncatus*), and thus a member of the whale family, sea mammals with which man has had no close contact until recently.

In Florida, a land of eternal summer, a new and unique town, Marineland, was founded 15 years or so ago, by a few inspired men, some 60 miles south of Jacksonville. The most important thing in this town that lies right on the seashore, is two aquaria of decidedly American dimensions, that is, the largest in the world. These gigantic pools are with justification not called aquaria, but oceanaria. One of these oceanaria is a good 10 ft. deep and 70 ft. across. From above as well as through about 200 windows like portholes, one can look into this huge tank from three different storeys or decks. Eleven six-foot porpoises play in it, and that in itself is a sight never before seen.

Feeding takes place three times a day. A sailor rings a submerged bell whereupon the strange heads of the intelligent porpoises appear above the surface. The animals wait for fish, which they catch skilfully, to be thrown to them from a sort of ship's rail. Finally these 450-lb. whales leap vertically out of the water to take the fish out of the hand, or even out of the mouth, of the sailor standing high above the water level ; but this is only a beginning. It turns into a fairy tale when a diver goes down into the water and feeds the porpoises swirling around him, as well as thousands of small fishes, out of an iron basket. The diver has the porpoises romping around him until he shows them the empty basket, whereupon they instantly wheel off and play with each other. Sometimes they emit their characteristic whistling, and beads of bubbles then rise from the valve-like breathing hole in the centre of the curved forehead.

Occasionally, one of the grey Florida pelicans, which are also at home in the circular oceanarium, loses a feather, and the porpoises play with it for hours. It is known, too, that they can communicate with each other by whistling, and anatomists have often pondered over their remarkably well developed brains, similar to those of apes. In ancient Greece, men thought very highly of the Mediterranean dolphins. It was thus fitting to undertake training experiments with these ever lively and playful mammals that so clearly need to be active.

At short notice, the management of the Marineland studios invited an experienced sea-lion tamer, and gave him the job of seeing what he could do with the porpoises. The result was surprising, not to say staggering. No porpoise had ever been trained before by anybody, nor could anyone predict what would be the outcome of the bold and original experiment.

The trainer chose out of the school in the oceanarium a young adult porpoise, " Flippy ", and transferred it to a circular tank 25 ft. in diameter and 4 ft. deep, away from the public part, where he could talk to his porpoise in private. I had the good fortune to be allowed to visit Flippy in his private practice tank. He dashed over to me and could not have enough stroking and scratching, especially on his light-coloured throat. The confidence of this legendary sea creature, its exaggerated human eyes, its strange breathing hole, the torpedo shape and colour of its body, the completely smooth and waxy texture of its skin and not least its four impressive rows of equally sharp teeth in its beak-like mouth, made the deepest impression on me. Up till then I had only seen such creatures as stiff, remote figures curling along the walls of museum entrance halls, or caught a glimpse of them from the deck of a liner as they approached in great schools, leaping fish-like over the waves. But Flippy was no fish, and when he looked at you with twinkling eyes from a distance of less than two feet, you had to stifle the question as to whether it was in fact an animal. So new, strange and extremely weird was this creature, that one was tempted to consider it as some kind of bewitched being. But the zoologist's brain kept on associating it with the cold fact, painful in this connection, that it was known to science by the dull name, *Tursiops truncatus*.

This, however, was no time for musing. The trainer, with the porpoise at his heels just like a dog with his master who will some time or other throw the stone for him to retrieve, had meanwhile stuck paper over a hoop of about 3ft. diameter, and hung it over the centre of the tank. Soon came the command " Jump " and Flippy jumped clean through the paper hoop, and was now putting his expressive head out of the water, looking for both applause and fish. Next it was the turn for another trick. " Back off ", came the order, " back to the far end of the tank ". The porpoise obeyed promptly. " Catch ", and Flippy skilfully caught between his saw-like rows of teeth the tennis ball thrown to him, and dashed over to his master, with one propeller beat of his horizontal tail fin, to exchange the ball he had retrieved for a piece of fish.

Next he was ordered to jump out of the water and pull a handle so that a bell rang. A stick, too, was thrown and caught with unerring assurance, and then brought back ; Flippy turned on his own horizontal axis in the water at a high speed at the word of command, or played a musical instrument by biting it rhythmically. He was even used as a draught animal ; equipped with a special harness, he was attached to a surfboat in the neighbouring lagoon and on the word go, drew bathing beauties over the blue water. An ancient Greek legend, human beings riding on dolphins in the water, had come to life in Marineland.

There are imperceptible gradations between playing and training, as the few examples quoted show ; these two things are not opposites. Good training is disciplined play. Both play and training often give excellent opportunities for brightening up the daily existence of the animals in the zoo, making it more significant, and giving the animal the necessary amount of exercise and occupation.

11

THE ANIMAL'S EXPRESSION

In zoological gardens, where animals from every class of the animal kingdom and from all parts of the globe come together, and in which they should lead pleasant lives under artificial conditions, man should learn far more than anywhere else about the most important ways in which his charges express themselves, and so be able to form a picture of their internal condition from these external signs. If anybody in a zoo mistakes the meaning of these phenomena of expression, he is heading for failure, or will immediately get a reminder in the shape of a surprise defensive action, for example.

That is why the greatest attention must be paid above all to the animals' expressions in the zoo, as these provide an important factor in determining the animals' mental and physical well-being. Characteristic changes in facial expression, and in the body, are of frequent occurrence in most animals.

The more practice one has, the more signs one recognizes in the animal and the greater the range of expression appears. Yet how easy it is for anyone unfamiliar with animals' expressions to be mistaken, can often be seen in the zoo. One of the most astonishing examples I can still remember vividly. While conducting a party of enthusiasts round, I went to see a recently caught marten, not on exhibit but kept for the time being locked up in a service room.

The timid creature had retreated as far as it could, that is, under an inverted sleeping box in a pile of hay it had dragged in. Apparently the marten only left its hiding place at night, when all was quiet, and nobody was about. To a newly caught animal, man always means a deadly enemy. I did not want to disappoint the members of the little party and, with some reluctance, uncovered the hidden marten by lifting up the box that covered it. The animal now lay completely exposed to a dozen staring people. One of them volunteered the remark : " Ah, it's asleep." Hopelessly wrong of course ! Far from being asleep, the marten was in a state of extreme excitement, not knowing, after it had recovered from the momentary shock, whether to jump at me or to make a dash for the farthest corner of its prison. In this case, sleep was a complete and utter misunderstanding—if anyone had gone too close to the so-called sleeping marten, he would have recognized a state of surprising wakefulness in the form of sharp teeth, and probably would never again have mistaken a sleeping marten for an extremely alert one !

Another instance of a completely topsy-turvy interpretation of an animal's expression was so outstanding that I made a note of it a couple of years ago. In those days, the original stags' house, built in North Scandinavian style, was still standing opposite the raptorial birds' enclosure. In one compartment we had temporarily put an African barbary sheep with its young lamb, and now we had to tackle the problem of getting these two wild sheep back to their flock in another part of the gardens. This seemed so easy and indeed would have been if it had been a question of domestic sheep. The movement of wild sheep, however, can be a ticklish problem, especially when there is no suitable equipment. Thus, for example, there were no drop-doors in the dilapidated old stall—since demolished—but only a useless swing gate, which precluded the possibility of barring in the animals for feeding. The gate was not in the only suitable spot, a corner of the enclosure, but in the centre of the narrowest side, and it projected somewhat into the interior of the enclosure. It was therefore practically impossible to drive the animal along the fence into the stall, since it dashed into the fatal blind corner between the fence and the door, where it stood in great excitement only to break desperately out once more into the middle of the enclosure. There was nothing left but to make a chain of keepers, hem the animal into a corner as quickly as possible, and, by means of a simultaneous grab of several men, to capture it by hand.

Anybody who has never watched a manoeuvre of this sort can have no idea of the agility and strength of a wild animal, especially a wild sheep. The maned sheep managed to break through the ranks repeatedly, forcing us to start hemming it into a corner all over again. Finally three men cornered the animal, which was as out of breath as the keepers, and were about to grasp it by the horns and feet when it turned like a flash and, instead of breaking out at the side, jumped straight into the face of the head keeper, who was standing in the centre. As might be expected, he ducked, but received such a kick on the back from the hind legs of the sheep as it broke out, that it knocked him flat on the ground, and at the same time he was grazed behind his right ear by a glancing blow from a hoof.

The animal was free again and dashed up and down the enclosure. At that moment a lady with a large party of children came on the scene, rummaged around in a paper bag, and, blissfully ignorant of what was going on, cheerfully held out a piece of bread to the foam-flecked maned sheep, as it dashed by her. Excited animals do not react to food stimuli ; one can only hope to calm them down with food at the very beginning of a state of upset.

What astonished and interested us was that this lady visitor to the zoo had completely failed to recognize the extraordinary state of excitement of the maned sheep. Let us consider, then, by what signs this could have been recognized. First of all, the total situation, quite apart from the disappointed keepers, panting for breath : the animal was not just running, it was dashing up and down, and along the fence as well. This had nothing to do with play, or desire for exercise ; the whole external appearance of its movements was obviously caused by some strong urge or other. Direction, tempo and type of movement of the animal were expressive of extreme distress and excitement.

In addition there were many other characteristics of expression, completely ignored by the lady visitor, and yet which it is the duty of anyone who has to deal with animals to know. Between the tip of the animal's nose and the end of its tail, there was nothing but evidence of a dangerous degree of excitement, which gave us the utmost trouble to reduce and overcome. The nostrils, for instance, were in this case fully distended. The mouth was gaping and foaming with the tongue hanging partly out. Ungulates differ from predatory animals or monkeys in not being able to open their mouths very wide, yet even the relatively narrow opening of this wild sheep's mouth was enough indication of extreme excitement. Its eyes were staring and starting out of their sockets, their expression alone was a typical symptom of upset. Additional signs were the violently heaving flanks and the hoarse panting—an audible form of expression. Its hair was bristling along its back, and its tail stood up vertically stiff, a characteristic sign. In this case, it is true, there was no special cry, as with many other animals ; but I think that the characteristics of expression mentioned were enough to tell any observer of average experience what was the internal state of the animal. I have gone into such detail over the case, as I wanted to show that there was a complete misunderstanding between an animal and a human being in the zoo, and this sort of thing unfortunately is commoner than one would suppose.

In spite of a sound knowledge of animals' lives and extensive practical experience, it constantly happens that animals' expressions are misunderstood by man, all the more so if the animal is remote from man in the zoological system, and thus the more dissimilar its construction.

So, until recently, " snakes' dances " have been described and illustrated in the herpetological literature, as examples of highly developed mating ceremonial. Davis (1936) and Stemmler (1935) among others, have described many highly ornamental postures struck by poisonous and non-poisonous snakes during their mating display—corkscrew interlacings or splendid, more attractive lyre-like shapes, etc.

Exhaustive investigations into the wealth of snake material in the terrarium at San Diego Zoo, California, have nevertheless led the herpetologist Charles E. Shaw (1948, 1951) to the conclusion that all these snake dances are to be regarded as fighting behaviour between rivals of the same sex and also to give proof of the correctness of his interpretations. What seemed even to experienced observers to be obviously mating behaviour has been shown on closer anlaysis to be the opposite. Thanks to Mr. Shaw's kindness, the Zoological Society of San Diego has allowed me to illustrate this book with some unique pictures of a fight of this kind between male *Crotalus*. I am most grateful to them for it.

Although animals, particularly wild ones, are as a rule excellent observers, far superior to men, it sometimes happens that they misunderstand one another ; in other words they misinterpret the expressions of other animals. In the zoo, where animals that are quite foreign to each other come into contact, this happens comparatively often (*cf.* 1950, pp. 111 ff).

I once watched a harmless example of this sort in " Artis ", the Amsterdam Zoo, when the director, Dr. A. F. J. Portielje, now retired, was showing off his tame condor. As soon as this giant vulture spied its human friend, it flew down to the floor of its spacious cage, and hopped over to greet its trusted human friend with outstretched wings. At the same moment, another vulture that until then had been sitting in an adjoining cage without taking any notice, turned in the same direction as the condor, and spread its wings out, exactly parallel to those of its neighbour. A few minutes later, however, this second vulture looked round puzzled, clapped its wings together and walked off again.

What had happened in this case ? The second vulture had obviously misunderstood a gesture, an expression on the part of the condor. It considered the spreading of the condor's wings as the expression of a desire to sun itself, and that sort of a mood is very catching among birds. Just like bathing, rest, and preening, sunning often occurs, especially among the larger birds, as a sudden collective state of mind. One begins by spreading out its wings so that the sun's rays catch it vertically ; a second follows, and in a twinkling all the others have also done so. In Africa, I watched hundreds of standing marabou storks—whole fields full of them so to speak—apparently turn black as they simultaneously opened their wings to sun themselves. The vulture in Amsterdam Zoo clearly mistook the condor's similar action of greeting, during which it spread its wings out wider. Soon afterwards, it realized its mistake and discovered that the sun was not shining at all.

Anyone critically inclined can of course object that this is just assertion without proof, since I could never know what was going on in the vulture's mind. The objection is theoretically justified. On the other hand, it must be stated here and now that people who have some experience in this field, and who see hundreds or thousands of animals daily eventually, by noting the expressions, develop a kind of intuition, rather like a doctor's diagnostic sense.

There is no lack of experiments for accurately recording the zoo animals' expressions. For example, Rölf Worner (1940) filmed the facial expressions of Rhesus monkeys. Then each separate picture was projected on to transparent tracing paper and copied, with all superfluous detail omitted. The pictures thus produced were laid on finely meshed graph paper, and the mobile portions of the face plotted at several points on this system of co-ordinates. In this way, interesting graphs could be made of the combination of separate mimic details, *e.g.*, the movements of the mouth, eyebrows and ears.

The Dutch physiologist and animal psychologist F. J. J. Buytendijk (1935) used much the same method of analysis with film strips to examine the reciprocal behaviour and forms of expression during a fight between a mongoose and a cobra. In the thesis by Rudolf Schenkel (1947) on studies in expression among wolves, many externally visible facial expressions, gestures, examples of bearing, ear movements, as well as a dozen various tail positions are shown in nearly a hundred separate pictures. In many animals, in fact, the tail is a particularly sensitive indicator of inner feelings. This extension of the vertebral column plays such an outstanding part not only in expression but in countless other ways, that Peter Bopp has undertaken to write a thesis in the zoo on the different functions of the tail.

If it was a question of simply defining and classifying the animal's expression according to the needs of zoo practice—for home use, as it were—without having to bother about theoretical claims, we might proceed as follows. By animal expression we would mean only those variable non-pathological phenomena of the animal which may help to an understanding of their situation. Here we must make it clear that it is not always easy to draw a definite line between pathological and non-pathological signs. According to one definition, vomiting, abnormal excrementation, and increased glandular secretion, may count as expression phenomena. For example, diarrhoea in a tropical ungulate may mean : a cold, parasitical attack, errors in diet, or excitement ; in practice, a swift differential diagnosis is not always easy (see below). In connection with this, the four following broad main groups of expression phenomena may be distinguished for vertebrate animals, illustrated simply by a few suggestive examples.

1. Acoustic.
2. Optic.
3. Olfactory.
4. Internal.

1. *Acoustic phenomena of expression (production of sounds)*.

In this category one may distinguish :

(*a*) *Vocal, i.e.* sounds produced through the larynx or syrinx. These include special cries of mammals and birds, the barking of dogs and seals, the purring of the larger cats, and the crying of martens.

(*b*) *Nasal, e.g.* the whistling of chamois, ibex, tahrs, blue sheep (*Pseudois*), barbary sheep (*Ammotragus lervia*), reedbuck, and marmots.

(*c*) *Dermal, i.e.* sounds produced by the skin, *e.g.* the rattling of the porcupines, *Oreotragus* and rattlesnakes ; the rubbing together of specially shaped scales as in *Echis*, and the gnashing of teeth in bears. Beak clapping, and rattling, noisy in the case of storks and gentle in toucans and owls, may all be included in this category.

(*d*) *Noises produced with the help of surrounding objects*. The classic example is the already-mentioned " splash-sound " of the frog caused by the impact of its body, under specific flight reaction, hitting the water surface with a characteristic noise, thereby warning its neighbours squatting on the bank. The swimming beaver makes a very similar, albeit louder, sound when he is alarmed in the water, and smacks his wedge-shaped tail hard on to the surface of the water, so that this danger signal carries a long way, and the water spurts up seven or eight feet. The rabbit's (*Oryctolagus*) thumping of its hind legs on the ground is equally characteristic. This danger signal may also be observed in the tame rabbit, when for instance a dog is set running near its hutch and startles it. Porcupines (*Hystrix*) also thump their hind legs noisily against the ground when alarmed. Camels stamp with their forefeet ; chamois and tahrs do the same with both feet simultaneously, when in a state of excitement. The well-known drumming of the woodpecker on decayed boughs, on the tin protective caps of telephone posts, or against rattling window panes belongs here, as well as the impatient banging on partition doors by the great cats or bears.

(*e*) *Other sounds*. These occur, for instance, when fish expel air through change in pressure, when brought up to the surface from the depths of their home waters, or when amphibia are removed from the water. As a territorial demarcation signal the male ostrich uses a lion-like rumbling roar, caused by the inflation of the oesophagus, at the same time closing the gullet above the stomach.

In the zoo, the sound of excrement falling on the ground may partake of the character of expression. This may have practical significance in the case of giraffes ; normally, the falling of faeces should give a typical rustling

Figure 16c.—One of the fighting pair attempts to push the other one's head away from beneath. Occasionally it slips and tilts over; both then let go and resume their fighting position.

Figure 16d.—Now the two fighters confront each other afresh, bringing their under-scales into contact at about the level of the heart, and pushing hard against each other, while their heads face each other, like mirror images, at the characteristic angle position of 45°. Simultaneously, both creatures perform sideways waving and swaying movements with the upper parts of their bodies.

Figure 16e.—*The fighting tactics seem to be aimed at overthrowing the opponent, and literally "flooring" him. This has clearly succeeded here. The victor is pressing so heavily on his victims head that the latter has difficulty in releasing itself and retreating.*

Figure 16f.—*As they push against each other, the two fighting snakes lean over until the right-hand one almost bangs its head on the ground. In these complicated disputes, as Ch. F. Shaw has pointed out, neither the rattle nor the poison fangs come into play. These belong apparently to other functional spheres, namely prey and enemy.*
(Figures 16 a–f from photos by G. E. Kirkpatrick, Zoological Society of San Diego.)

sound.   If the excrement is voided in shapeless, pattering portions, this is an important guide to the keeper.

2. *Optical Phemomena of Expression.*
These may be roughly divided into facial, gesture, and colour-change phenomena.

(*a*) *Facial.*   By this we mean characteristic changes in the facial region, occurring on the physiognomy, the solid architecture, so to speak, of the face.   This includes emotionally conditioned changes in :

   i. *Ear position* (*e.g.* in cats, equidae, elephants, red deer).
  ii. *Jaw and lip position* (*e.g.* in bears, rhinoceroses).
 iii. *Whiskers* (*e.g.* in cats, dogs, sea-lions).
 iv. *Crest feathers* (*e.g.* in cockatoos).
  v. *Eye opening*, pupils, eye position (*e.g.* cats, anthropoid apes, parrots, herons).
 vi. *Tongue* (*e.g.* in snakes).
vii. *Nose* (tapir, elephant, seals).
viii. *Yawning*, often thought of as an expression of sleepiness, has however quite a different meaning with many animals (hippopotami, monkeys), and is to be interpreted as a danger signal (so-called " temper " yawning).

(*b*) *Gesture.*   This includes especially expression phenomena apart from the facial area, that is, on the body, the limbs and the tail.   Two sub-groups may be differentiated ;   namely static phenomena, which occur as it were at one spot, without change of place, and dynamic, which comprise those which do change position.

i. *Static*
General bearing (erect with stilted legs, bent, ready to jump, ready to withdraw, *etc.*).   In the case of snakes, the observer is almost exclusively concerned with general bearing; pupils, tongue and sounds may give some help.
Bristling of the mane (*e.g.* giraffes, wild boars, canine family, maned bovidae, rodents).
Tail movements (*e.g.* the twitching of the tip of the tail in cats, horizontally and vertically in canidae, erection in ostriches, tucking in and erection in antelopes and canidae).
Ruffling of feathers (in owls, peacocks, pheasants, *etc.*).
Display of rump patch (roedeer, antelopes and red deer).
Stamping (chamois, tahr, maned sheep, horned sheep, mouflon, camelidae, *etc.*).

ii. *Dynamic*
Tempo of movement forward.
Kind of step (*e.g.* goose-stepping in deer).   The kind of step has also proved to have important expressive value in man (G. Kietz, 1952) ;   similar studies for animals would be most interesting.

(*c*) *Change in colour.*   The colour change in cephalopods as an expression of internal state is famous.   Blushing and turning pale in man—even though this does not depend on the contraction and dilation of

chromatophores as in the octopus—are included among those phenomena. The chameleon is the classic example ; its change of colour is
less the result of camouflage than of mood. Many fishes are in no way
inferior to the famous reptile in this respect. Even bare patches of skin
on birds, and the ears of the Tasmanian devil (*Sarcophilus*), for instance,
have quite different coloration, according to the emotional situation.

3. *Olfactory expression phenomena.*

These, too, occur in many forms, even in *Homo sapiens*. The lie detector,
as is well known, depends upon variations in the electrical conductivity of the
skin according to the activity of the sweat glands. For man, a predominantly
optically orientated creature, it is used to detect by eye, with the aid of a
galvanometer, these variations in the glandular activity. On the other hand,
macrosmatic mammals, *e.g.* the dog, are able to interpret directly through the
nose, thanks to their literally superhuman sense of smell, such changes in the
functioning of the glands without having to transform them optically.
Thus the dog has rightly been described as "microlfac" (R. and R. Menzel,
1930, p. 170), and many of the dog's apparently mysterious reactions to man's
emotional state or illness are to be ascribed to this capacity. Many diseases,
that are accompanied by characteristic changes in the scent formation, might
thus be diagnosed very early with the help of a dog (Katz, 1948, p. 72).

In addition to the optically perceptible, there are also mimicry and physiognomy of smell, usually accessible only to macrosmatic animals. In
exceptional cases, men in whom the sense of smell has been particularly well
cultivated, find it possible to detect scent mimicry in animals, *e.g.* mice
(1946).

Functional changes in those animals with very strong skin glands are
clearly perceptible even for humans provided with an average sense of smell.
Many animals have definite stink glands, *e.g.* many prosimia, stoats, polecats
and their kind. The skunk (*Mephitis*) is an extreme case, actually using its
glands for defence and, under critical reaction, shooting a real poison gas,
namely butylmercaptan, at its opponent.

The counterpart of the skunk among birds is the hoopoe (*Upupa epops*).
While in the nest, the nestlings and the female are able to eject their excrement with fair accuracy, at the same time producing from their perineal
glands a most offensive secretion (E. Sutter, 1946). Bats can produce from
their facial glands a stinking secretion, when threatened. I have no space
to detail the numerous examples of offensive glandular scent changes in the
various species ; I simply draw attention to J. Schaffer's comprehensive
work (1940) in this connection.

4. *Internal expression phenomena.*

Here, I am not so much thinking of the secretion of adrenalin, of the
traumatic Basedow disease in rabbits, *etc.*, but principally of the reactions of
the digestive tract (in the widest sense) to disturbances, such as we often see
in the zoo, or during the capture of animals. Here one is tempted to think
of the well known phenomenon in human beings, when in certain emotionally
caused situations one's tongue sticks to the roof of one's mouth, *i.e.* the
typical reaction of the salivary glands.

Anyone who handles snakes knows how carefully they must be treated after feeding (voluntary or forcible) to prevent them ejecting the whole of the contents of their stomach. The same thing happens with many lizards, *e.g.* monitors. The slightest disturbance can lead to regurgitation. This stomach sensitivity can also be observed in other Sauropsidae, in birds, especially in those which do not fill their crops with fine particles of food but gulp down large chunks into their stomachs or gullets, as is the case with fish-eaters (pelicans, ibis, and herons). Many raptorial birds also vomit up the contents of their stomachs. In the case of many birds, this may help them to escape from enemies by allowing swifter, unencumbered flight. Small seed and insect eaters do not show this phenomenon.

Naturally we include here the emptying of the bladder in conditions of excitement, *e.g.* in elephants. Great bats that have been frightened use their urine as a weapon of defence as well, sprinkling their enemies with it with astonishing precision. Toads and tortoises often manage to gain their freedom for a moment through a sudden surprise emission of urine, when they are being unnecessarily handled by human beings. Many amphibia react to rough handling by emitting poisonous skin gland secretions.

During the daily zoo inspection, the zoo manager has to keep an eye open for a thousand and one possible indications of expression. Not only his eye, but his ear and even his nose must play an active part. There may be many discords in the thousandfold chorus of sounds, from the peacock's warning screech to the porcupine's rustling, the Himalayan tahr's whistling to the emu's trumpeting. Frequently the nose of the man who looks after animals recognizes particular smells, which, as the outcome of a change in gland activity, may act as important symptoms of internal conditions. In prosimia, many of the smaller predators, and rodents, a quick change in the scent formation according to the mood of the moment may be observed. Not everyone responsible for animals goes so far in the interpretation of such scent signs as some experts on mice, who believe they can identify surprising variations in the internal state of their tiny charges through the change of body smells.

The meaning of internal expression phenomena is evident to the observer in the zoo, not only in striking variations in appetite but in a really drastic way when, for instance, he walks unexpectedly into the heron's cage shortly after feeding time. These fish-eating birds express their dislike of that sort of disturbance by emptying their stomachs, and sending a regular shower of fish down on the human intruder. Pelicans, too, react to the least disturbance by completely emptying their stomachs ; the business of cleaning out their pond, or repair work in the vicinity of their cage, is enough to cause this.

Refusal of food comes into the category of internal expression phenomena in zoo practice. When, for instance, emus are transferred from one cage to another, fasting for weeks afterwards may result. The elephant is well known for the sensitivity of its intestines. The most trivial thing may cause a change in functioning, such as the flight diarrhoea in elephants in the wild, known to all elephant hunters. The slight disturbance connected with crossing over a busy motor road in Africa may cause a watery evacuation of the bowels in an elephant.

Camels, too, are sensitive in this respect. A camel foal that we were training to walk through the town responded to every forcible walk outside its familiar enclosure with increasing diarrhoea. The further it got from its well known stable, the faster did its originally firm round droppings turn to watery ejections, until eventually, after carefully graduated practice walks, the disappearance of this significant wateriness, together with other symptoms of course, proved to us that the camel no longer objected to short excursions outside its enclosure.

Every one who looks after anthropoid apes knows that these popular show animals possess a very expressive digestive canal. Even slight scolding may entail a great increase in cleaning-up operations.

Naturally, highly developed mammals in particular betray by their facial play, their cries, their tail movements, and so on, a wealth of expressive phenomena which must be noticed by even the outsider. In mammals, the whole body is like an open book, to be read by those who know how. Every item, from the way the hair lies, to the position of the tip of the tail, has its special meaning. As a rule, the closer one gets to an animal, the clearer one finds the expressions of its condition, i.e. the external symptoms of the mood. On the other hand, there is a large number of expression phenomena we can understand the first time we meet, as they are more or less non-specific or resemble those with which we are already familiar in other animals. Indeed, some human grimaces are found in much the same form in animals.

There are, of course, animals which are markedly lacking in powers of expression. Thus reptiles and most birds have very little facial expression, or none at all. Among the larger predators, bears, especially polar bears, are feared, because one cannot tell from their faces what they are up to. That is why trainers have such a hard task with these animals. In the case of completely strange species of animals which are encountered for the first time, one may be completely at a loss to start with, but even with the lower vertebrates, a number of facts appear on closer inspection, which may help in determining the state of mind of a fish, a salamander or a tortoise. In fishes, for instance, a very good indication of its internal condition is given by the way the fish moves about in the tank. Madly brushing along the sides of the tank and swimming up and down in a corner are unmistakable signs of discomfort. A fish that is used to swimming around its tank, just as a bird that knows its aviary, never touches the boundaries of its living space but shows by its elegant sweeping movements its awareness of those limits. Other aids are the fish's change in colour, and the action of its fins—often surprisingly accurate and precise indicators of internal state of affairs. There are fish, such as our common native minnows, which may show a distinct colour change on escaping. Many fish have very mobile eyeballs, and can thus roll their eyes.

Moreover, the fact that the most important facial nerve in human beings, the nervus facialis, in this branch of the animal kingdom, controls a considerable region, that of the gill-covers, shows how wrong it is to regard these denizens of the water as expressionless. These very gill-covers, with their degree of splay and the rhythm of their movements, are highly expressive organs to the student of fish. Here too, as with many mammals, there appears an overlapping of functions in many cases, and this frequently leads to mistaken interpretations of the animal's expression.

In the African elephant, for example, the ears, that is, the giant ear flaps, may be highly important expression organs ; at the same time they are useful regulators of the body temperature. At certain degrees of temperature and humidity, these organs, with their large area and good blood supply, move according to certain rules. The movements thus set up may be confused with purely expressive movements, and deceive the observer. In deer, the independently movable ears are used for detecting sounds, but may suddenly, when turned as far back and down as possible, form the part of a facial expression, with the meaning of complete readiness for battle. In climbing fish, the splaying of the gill-covers, otherwise a sign of defence or disturbance, may, in certain situations, sometimes be used to help the fish to creep along. Often the gill-covers of fish are provided with a variety of mechanical or chemical weapons in the shape of spines and stings. With them, the splaying of this organ is not just an " empty " gesture, but an actual preparation for fighting, a final warning, a threatening gesture.

It is very often difficult to know the proper meaning of breathing movements. Rapid pulse and breathing in man may sometimes, as we know, be a very significant symptom, or simply the result of physical activity. In many amphibia, breathing movements are hard to interpret. Olga Leffler was the first to point out, in 1914, that the frequency of gill movements in the axolotl depends to a certain extent on its nervous condition. In this familiar salamander-like amphibian from Mexico, nowadays a universal laboratory animal, the external gills project behind the head like delicate tufted fans, and move with a definite rhythm to and fro. Naturally, this movement depends on water temperature, oxygen content, *etc.*, and on state of mind, too. If an axolotl is gently stimulated, the number of gill movements per minute can be increased threefold.

In amphibia without external gills, the movements of the skin of the throat, the so-called gular oscillations, are important measures of the degree of excitement, as they are in turtles, particularly the aquatic species. For the rest, the reptiles are as extraordinarily varied in their phenomena of expression, as they are in their external appearance. The crocodiles, and their relatives, present the greatest difficulties to the observer. It is of course a completely mistaken and inadmissible anthropomorphism to say that these armoured lizards wear a constant smile on their faces, just because the corners of their mouths are turned upwards between their powerful jaws. We really have very little to go on to help us to guess even approximately the internal state of a crocodile. The few reliable indications include the pupils, which may be almost completely circular not only in the dark but also in states of excitement ; they become extremely thin vertical slits when the animal is in a calm and settled mood. This poverty of expression, which is so alarming in zoo practice, is the reason why one is not prepared for the sudden explosive reactions of crocodiles, interrupting their states of stony rigidity.

Some snakes may show more signs of expression, even though they have no external arms and legs. Vibratory tail movements among rattlesnakes, as well as many other poisonous New World snakes, are valuable danger signals.

The darting tongue, the bellows-like respiration, frequently accompanied by hissing, and the manner of twining are full of significance. In snakes, the whole body has become an organ of expression. Everyone who keeps snakes knows that it is the front third of the total body length that needs one's attention. If this part is formed into an S-shape, like a coiled spring, so that at any moment it can be shot surprisingly far forward, one must act with all possible caution ; if, however, the front part of the snake's body is stretched out flat, this is a sign of harmless confidence.

We find among the higher mammals the greatest wealth of expression phenomena. Not all of them, however, are so easy to read as the lion, with its almost exaggerated grimaces and cries. I always feel that he is trying to give human beings, who are rather slow in the uptake, elementary lessons in the meaning of animal expressions. Yet his nearest relations, leopards, pumas, and tigers, make it much less easy for us.

Charles Darwin pointed out the many similarities and even correspondences in his far-reaching work " Expression of the Emotions in Man and Animals ". One important difference between man and animals, however, lies in the fact that conscious deception through shamming, *i.e.* falsified expression, really occurs in man alone. The pretence of lameness, in the crane or golden plover, the famous shamming dead of the opossum, or the often doubted but really genuine shamming sleep of the fox has certainly nothing to do with deception. We look for this characteristic in vain in the animal kingdom, a fact which has earned for animals many friendships with human beings. Exceptions occur only among trained animals, whose characters, as D. Katz (1927) expresses it, have been " tainted by man's fall from grace ", and occasionally among very highly developed species, such as anthropoid apes living in close contact with man.

An example of a chimpanzee of this sort has already been referred to in the chapter on Play and Training. Another very illuminating one was that of the Gorilla " Achille " in Basle Zoo that employed all his cunning to entice people into his cage in order to satisfy his overpowering need for society. It is not difficult to gain the friendship of young apes ; it is much more difficult to rid oneself of these faithful creatures after a time.

One of the countless tricks that the four-year old gorilla used for procuring human contact was to push its arm out through the top of the wire mesh of his air-conditioned cage and pretend that he couldn't get it back again. Several times Head Keeper Carl Stemmler, before he realized that it was all a humbug to try to get some human company, hurried to help the gorilla out of its plight.

As these tricks had been seen through by the whole staff of keepers, it was by a queer stroke of fate that a new and quite inexperienced assistant woman keeper was working in the bird house, where the ape's air-conditioned cage was. A particularly unfortunate chain of circumstances brought it about that one day, at closing time, the new assistant was left alone in the house after the head keeper and the other keepers had already gone.

As the girl was looking around just before going home, her glance fell upon " Achille ", who, instead of lying on its wooden bed, was hanging in obvious despair high up from the cage bars and struggling unsuccessfully to free its arm from them. Not surprisingly, the young assistant felt sorry for the poor creature, and wanted to get it out of its plight. Instead of phoning me to report the incident, she thought she would effect the rescue by herself, took out the key of the gorilla's cage, and stepped inside to go to its assistance. The sly gorilla must have foreseen this humane action, for it was waiting, just behind the door, for the assistant. No sooner was this opened than the young ape, weighing a good seventy pounds, jumped up and clasped the assistant round the neck before she got inside the cage, forcing her back under the impact of this overwhelming encounter and down the three steps into the outer room, the food store with which the gorilla was familiar.

The new assistant naturally had only one thought—to get her precious charge back into his cage again. This she managed to do ; she slammed the cage door behind her, but in the heat of the struggle, she had dropped the key in the outer room. So there she was, sitting in the air-conditioned cage with the gorilla, still " lovingly " clinging round her neck. By a malicious twist of fate, her absence was not noticed at the check-out point, where she should have made her daily report.

It was not until the next morning that it was learnt, to everybody's horror, that the poor, but fortunately strapping girl, had had to spend the whole night in the gorilla's cage, clasped all the time by the gorilla that pined for somebody's company !

From time to time, when the gorilla dozed off, she tried to loosen its tender, but none the less close, embrace. The little gorilla would then regularly wake up, and cling even more closely to its foster mother. Next morning we found the animal-loving lady keeper, who unfortunately had not been missed at home overnight, in relatively high spirits, although completely wet through, and not very presentable, and we were able to release her from a situation one does not get into every day.

It is unnecessary to add that the press played up this incident for all they were worth : the girl became a Hollywood lovely, the young gorilla a full-grown monster, and in the next few days a number of visitors came to the zoo on purpose to view the girl concerned and the gorilla. About eighteen months later we learned to our great surprise that the baby gorilla, Achilles, now famous, was a female. In this species of anthropoid apes, sex determination is notoriously difficult, and in the case of young animals practically impossible (Hediger, 1952). In animals that have gained special experience through intimate contact with man, it not only happens that they sometimes practise shamming on human beings, as the examples quoted about apes show, but also that they can interpret human expressions expertly. Thus I remember a certain Airedale terrier that had little opportunity of frisking about because it was shut in a small front garden. This was about four feet or so above street level and was separated from it by iron railings. I often used to walk past this garden on my way to town, especially when going to give a lecture and consequently rather preoccupied. It often happened that the dog, obviously bored, suddenly barked at me from about shoulder height and gave me a thorough fright—to its obvious amusement. Significantly enough, this unexpected barking only occurred when I was deep in thought.

If the dog was in the street, for a change, it never thought of barking as I walked past. Similarly it refrained from barking suddenly when I was not preoccupied, and let it understand that I had already seen it by giving a gentle call as I passed. Sooner or later it would happen that I did not think about the dog until too late, whereupon it again played its little trick and startled me.

Simulation is thus apparently not impossible in animals that have been " spoilt " by close contact with man, *i.e.* trained or domesticated. Thus for example E. Frauchiger (1945, p. 193) describes dogs that pretended to be weak or lame when they had to work ; " but when they thought nobody was looking, they dashed about playing with bones in the kennels or racing after their kennel companions ".

The interpretation of animals' expressions is not always easy, as the examples quoted earlier have shown. It is thus no accident that the greatest blunders of this century in animal psychology can be ascribed, in the last analysis, to misinterpretation of animals' expressions, for instance the catch words " marsupialian stupidity " and the historic error concerning thinking horses and dogs—the Clever Hans error as D. Katz called it.

It is hard to understand nowadays how, less than forty years ago, distinguished and worthy people as well as scientists could have fallen victims to such a gross delusion. A few lines from the published works of two well known Basle personalities suffice to illustrate this remarkable aberration.

Professor Gustav Wolff, a psychiatrist (1914, p. 457) : " Once more a great and revolutionary event has occurred, outside all organized science. Like everything else that is new, it has to fight against the dogmatic attitude of scholars and the church. Suspicions are being repeatedly voiced ; but the work accomplished is so great that genuine doubts among those who seek the truth can only contribute to its fulfilment, while incompetence and malevolence can no longer harm it. Whoever has been to Elberfeld and Mannheim, and seen the miracle with an open mind and without prejudice, knows that animals can think like human beings, and can express human thoughts in human language—I do not know if I should have believed thus had I not myself experienced it with, I might almost say, awe."

Dr. Paul Sarasin, the well known naturalist and spiritual father of the Swiss National Park, says in the *Transactions of the Society of Natural History of Basle* (1915, p. 71) : " Our eyes have been almost dazzled by the results of the new science of animal pedagogy, and they must first get accustomed to the light that shines forth from the abundant observations for which we are indebted to the unflinching efforts of Karl Krall and Frau Paula Mökel, as well as to the important inferences for our conception of the world arising therefrom ; but when we eventually succeed in doing so, however great the astonishment we feel at the Mökel results, however much we are taken aback or perplexed, we still are unable to admit that in one particular intellectual activity, that is, in working out difficult arithmetical problems in their heads, the animals mentioned, especially the horse, could be superior to us, that is, to the average man. Yet it is unquestionably established that Herr Krall's stallion ' Muhamed ' in particular gives the right answers out of its head to sums involving square roots, which few of us could solve in a short time, while the right answer to the problem is always forthcoming with disconcerting speed. I have already pointed this out as a problem in my first report (1912, p. 252), and I draw attention to the fact that I emphasized this as a problem, in order to establish at the outset that all the critics were wrong to think that we, who were already convinced after repeated personal investigation into the justice of the Krall-Mökel claims, had not realized the seriousness of the problem involved. On the contrary, we knew it only too well, at least as well as Krall's critics, if not better. This is not said in order to boast, but to forestall stupid attacks."

We know today, as we have already said, that these apparently alarming performances by thinking dogs and horses were pure illusion, and that the authorities who fell for them were guilty of unsurpassed naiveté with regard to expression phenomena, not only on the part of the animal, but of the human beings concerned. Today there exists a wealth of literature on this subject, which we cannot even hint at here, unfortunately, but must confine ourselves to one fundamental observation.

The most serious sources of error were discovered by a variety of authorities, including O. Pfungst, who died much too young. Above all the fact was established that certain actions of the talking animal could be attributed to definite involuntary signs.

In the animal kingdom, and among mammals in particular, there is an extremely widespread and remarkably highly developed faculty of interpreting human expression, usually with great accuracy. One might expect the domestic animal, which has been so intimately connected with man for centuries to be able to understand and act upon man's signs better than the wild animal. Animals, especially domestic ones perhaps, are better observers and more accurate interpreters of expression than men, if we exclude his technical aids such as films, microscopes, psycho-galvanometers, *etc.* When animals and men meet, it is the rule therefore that the animal can learn more about man from his expression than man about the animal, provided the higher animals and man without any special equipment are concerned. Many animals in fact are equipped with literally superhuman sense organs and superior strength and shortness of reaction time.

If that were not the case, when capturing animals or in running zoos, there would not be the continual surprises that are the result of a real superiority of the animal in certain fields of sensory activity and bodily organization.

BIBLIOGRAPHY

Alverdes, F. (1925). Tiersoziologie. Leipzig
Andrae, K. R. (1935). Die " Imu "-Krankheit der Ainu-Frauen. Ciba-Zs., Basel, No. 20, pp. 697–698
Antonius, O. (1937). Über Herdenbildung und Paarungseigentümlichkeiten der Einhufer. Zs. Tierpsychol., Vol. 1, pp. 259–289
Antonius, O. (1938). Zum Domestikationsproblem. Zs. f. Tierpsychol., Vol. 2, pp. 296–302
Antonius, O. (1951). Die Tigerpferde. Frankfurt a. M.
Asdell, S. A. (1946). Patterns of Mammalian Reproduction. Ithaca, N.Y. and London
Austen, E. E. (1907). Behaviour of Toads when confronted by a Snake. Spolia Zeylanica, Vol. 5, p. 32
Baege, M. H. (1928). Naturgeschichte des Traumes. Leipzig
Baerends, G. P. (1941). Fortpflanzungsverhalten und Orientierung der Grabwespe Ammophila campestris Jur. Tijdschr. Entomol. 84. pp. 68–275
Bally, G. (1945). Vom Ursprung und von den Grenzen der Freiheit. Eine Deutung des Spiels bei Tier und Mensch. Basle
Barbour, Th. (1934). Reptiles and Amphibians. Their habits and adaptations. Boston and New York
Bavink, B. (1949). Ergebnisse und Probleme der Naturwissenschaften. Zürich
Benchley, B. J. (1942). My Life in a man-made Jungle. London
Benchley, B. J. (1942a). My Friends, the Apes. Boston
Benedict, F. G. (1936). The Physiology of the Elephant. Washington D.C.
Beninde, J. (1937). Zur Naturgeschichte des Rothirsches. Leipzig
Berger, A. (1910). In Afrikas Wildkammern. Berlin
Bierens de Haan, J. A. (1929). Animal Psychology for Biologists. London
Bierens de Haan, J. A. (1937). Labyrinth und Umweg. Ein Kapitel aus der Tierpsychologie. Leiden
Bierens de Haan, J. A. (1940). Die tierischen Instinkte und ihr Umbau durch Erfahrung. Eine Einführung in die allgemeine Tierpsychologie. Leiden
Billard, G. and Dodel, P. (1922). Les Mœurs des animaux en rapport avec la disposition des yeux et la forme des pupilles. Cptes. rend. Séances Soc. Biol., Bordeaux. Paris. pp. 153–154
Bilz, R. (1940). Pars pro toto. Ein Beitrag zur Pathologie menschlicher Affekte und Organfunktionen. Leipzig
Bopp, P. (1954). Schwanzfunktionen bei Wirbeltieren. Rev. Suisse Zool. Vol. 1.
Boulenger, E. G. (1934). The Aquarium Book. New York
Bramstedt, F. (1935). Dressurversuche mit Paramaecium caudatum und Stylonychia mytilus. Zs. vgl. Physiol., Vol. 22, pp. 490–516
Brecher, G. A. (1932). Die Entstehung und die biologische Bedeutung der subjektiven Zeitheinheit des Momentes. Zs. vgl. Physiol., Vol. 18
Brock, F. (1934). Jahrmarktsdressur wilder Mäuse als Grundlage einer wissenschaftlichen Verhaltensanalyse. Verh. deutsch. zool. Ges., pp. 235–246
Bruhin, H. (1953). Zur Biologie der Stirnaufsätze bei Huftieren. Physiol. Comp. et Oecol. Vol. 3, pp. 63–127
Buck, J. B. (1938). Synchronous rhythmic flashing of fireflies. The quarterl. rev. Biol. Vol. 13. Baltimore
Bühler, K. (1922). Die geistige Entwicklung des Kindes. Jena
Burton, R. G. (1931). A Book of Man-eaters. London
Buytendijk, F. J. J. (1933). Wesen und Sinn des Spiels. Berlin
Buytendijk, F. J. J. (1935). Reaktionszeit und Schlagfertigkeit. Kassel
Buytendijk, F. J. J. and Plessner, H. (1935). Die physiologische Erklärung des Verhaltens. Eine Kritik an der Theorie Pawlows. Acta Biotheor. Vol. 1, pp. 151–172
Carpenter, C. R. (1934). A Field Study of the Behaviour and Social Relations of Howling Monkeys. Compar. Psychol. Monogr. Vol. 10, No. 2. Baltimore
Carpenter, C. R. (1940). A Field Study in Siam of the Behaviour and Social Relations of the Gibbon (Hylobates lar). Compar. Psychol. Monogr. Vol. 16, No. 5. Baltimore
Carpenter, C. R. (1942). Sexual Behaviour of free ranging Rhesus Monkeys (Macaca mulatta). J. comp. Psychol. Vol. 33, pp. 113–162
Carpenter, C. R. (1942a). Societies of Monkeys and Apes. Biol. Symposia. Vol. 8, pp. 177–204

Carus, C. G. (1866). Vergleichende Psychologie oder Geschichte der Seele in der Reihen-folge der Thierwelt. Vienna.

Curran, C. H. and Kauffeld, C. (1937). Snakes and their ways. New York and London

Dahl, Fr. (1922). Vergleichende Psychologie. Jena

Darling, F. F. (1937). A Herd of Red Deer. Oxford University Press

Darling, F. F. (1952). Social Life in Ungulates. Collintern. Centre nat. rech. sci. Paris. Vol. 34. Structure et Physiologie des Sociétiés Animales. pp. 221–226

Dathe, H. (1934). Eine neue Beobachtung des Känguruhgeburtsaktes. D. Zoolog. Garten N. F. Vol. 7

Davis, D. D. (1936). Courtship and Mating Behaviour in Snakes. Zool. Ser. of Field Mus. Nat. Hist. Chicago. Vol. 20, pp. 257–290

Deegener, P. (1918). Die Formen der Vergesellschaftung im Tierreiche. Leipzig

Dijkgraaf, S. and Verheijen, F. J. (1950). Neue Versuche über das Tonunterscheidungs-vermögen der Elritze. Zs. vgl. Physiol. Vol. 32, pp. 248–256

Ditmars, R. L. and Greenhall, A. M. (1936). The Vampire Bat. Smithsonian Report for 1936. pp. 277–296. Washington D.C.

Ditmars, R. L. and Bridges, W. (1937). Wild Animal World. New York and London

Eickhoff, W. (1949). Schilddrüse und Basedow. Beiträge zur Histo-Morphologie und Funktion der Schilddrüse verschiedener freilebender Tiere. Stuttgart

Einarsen, A. S. (1948). The Pronghorn Antelope and its Management. Washington D.C.

Erhard, H. (1924). Rätselhafte Sinnesempfindungen bei Tieren. Natur und Technik. Vol. 5

Ferrier, A. J. (1947). The care and management of Elephants in Burma. London

Fischel, W. (1938). Psyche und Leistung der Tiere. Berlin

Frauchiger, E. (1945). Seelische Erkrankungen bei Mensch und Tier. Bern

Frechkop, S. (1946). Notes sur les Mammifères. XXIX. De l'Okapi et des affinités des Giraffidés avec les Antilopes. Bull. Mus. roy. Hist. nat. Belg. Vol. 22, pp. 1–28

Frisch, K. von (1938). Zur Psychologie des Fischschwarmes. Die Naturwiss. 26. Jahrgang, pp. 601–606

Frisch, K. von (1948). Aus dem Leben der Bienen. Vienna

Garretson, M. S. (1938). The American Bison. New York

Glees, P. and Cole, J. (1950). Recovery of skilled motor functions after small repeated lesions of motor cortex in Macaque. J. Neurophysiol. Vol. 13, pp. 137–148

Goethe, F. (1939). Aus dem Jugendleben des Muffelwildes. D. Zoolog. Garten N. F. Vol. 11, pp. 1–22

Grabowski, U. (1939). Experimentelle Untersuchungen über das angebliche Lernvermögen von Paramaecium. Zs. f. Tierpsychol. Vol. 2, pp. 265–282

Green, E. E. (1906). Curious action of a Toad when confronted by a Snake. Spolia Zeylanica. Vol. 3, p. 196

Greppin, L. (1911). Naturwissenschaftliche Betrachtungen über die geistigen Fähigkeiten des Menschen und der Tiere. Biol. Zentralbl., Vol. 31

Gyr, W. (1946). Die Kuhkämpfe im Val d'Anniviers. Schweiz. Ges. Volkskunde. Basel

Hafen, G. (1935). Zur Psychologie der Dressurversuche. Zs. vgl. Physiol., Vol. 22, pp. 192–220

Hamilton, W. J. (1939). American Mammals. New York and London

Harms, J. W. (1948). Zoobiologie für Mediziner und Landwirte. Jena

Hartman, C. G. (1952). Possums. Austin, Texas

Hayes, C. (1951). The Ape in our House. New York

Heck, H. (1940). Elefantenbullen. Das Tier und wir. Munich

Heck, L. (1930). Aus der Wildnis in den Zoo. Berlin

Hediger, H. (1934). Zur Biologie und Psychologie der Flucht bei Tieren. Biol. Zentralbl. Vol. 54, pp. 21–40

Hediger, H. (1935). Zur Biologie und Psychologie der Zahmheit. Arch. ges. Psychol. Vol. 93, pp. 135–188

Hediger, H. (1935). Zirkusdressuren und Tierpsychologie. Rev. Suisse Zool. Vol. 42, pp. 389–394

Hediger, H. (1937). Die Bedeutung der Flucht im Leben des Tieres und in der Beurteilung tierischen Verhaltens im Experiment. Die Naturwiss. Vol. 25, pp. 185–188

Hediger, H. (1938). Ergebnisse tierpsychologischer Forschung im Zirkus. Die Naturwiss. Vol. 26, pp. 242–252

Hediger, H. (1938). Tierpsychologie und Haustierforschung. Zs. Tierpsych. Vol. 2, pp. 29–46

Hediger, H. (1941). Biologische Gesetzmässigkeiten im Verhalten von Wirbeltieren. Mitt. Naturf. Ges. Bern (1940)

Hediger, H. (1946). Mäuse und Menschen. Schweiz. Annalen, Vol. 3, pp. 103–106

Hediger, H. (1946). Zur psychologischen Bedeutung des Hirschgeweihs. Verh. Schweiz. Naturf. Ges. Zürich, pp. 162–163

Hediger, H. (1950). La Capture des éléphants au Parc National de la Garamba. Bull. Inst. Royal Col. Belge. Vol. 21, pp. 218–226

Hediger, H. (1950). Wild animals in captivity. An outline of the biology of Zoological Gardens. London

Hediger, H. (1951). Jagdzoologie—auch für Nichtjäger. Basle

Hediger, H. (1951a). Observations sur la Psychologie Animale dans les Parcs Nationaux du Congo Belge. Institut des Parcs Nationaux du Congo Belge. Brussells

Hediger, H. (1951b). Grundsätzliches zum tierpsychologischen Test. Ciba Zs. No. 125

Hediger, H. (1952). Seltene tropische Tiere und ihre Haltung in Zoologischen Gärten Nordamerikas. Acta Tropica, Vol. 9, pp. 97–124

Hediger, H. (1952) Dressurversuche mit Delphinen. Zs. f. Tierpsych. Vol. 9, pp. 321–328

Heinroth, O. (1933). Zur " Akinese " bei freilebenden Völgeln. Ornithol. Monatsber. Vol. 41, No. 5, Berlin

Heinroth, O. (1938). Aus dem Leben der Vögel. Verständl. Wiss. Vol. 32, Berlin

Heller, E. (1930). The American Prong-Horned Antelope. Bull. Washington Park Zool. Soc. Vol. 1, No. 4

Hempelmann, F. (1926). Tierpsychologie vom Standpunkte des Biologen. Leipzig

Heymann, K. (1943). Seelische Frühformen. Beiträge zur Psychologie der Kindheit. Psychol. Praxis Basel. Vol. 1.

Hill, W. C. O. (1952). Habits of the mothers and the young of Primates. Zoo Life, London. Vol. 7, No. 4

Hindle, E. (1947). The Golden Hamster. The UFAW Handbook on the Care and Management of Laboratory Animals. London

Hinsche, G. (1928). Kampfreaktionen bei einheimischen Anuren. Biol. Zentralbl. Vol. 48, pp. 577–617

Hinsche, G. (1939). Über die Entwicklung von Haltungs- und Bewegungsreaktionen. W. Roux, Arch. Entwicklungsmech. Vol. 139, pp. 724–731

Hinsche, G. (1944). Zur Genese der Stereotypien und Manieren. I. Wegeriten. Psychiatr.-neurol. Wschr. 46. Jahrgang

Hodge, W. H. (1946). Camels of the Clouds. The nat. geogr. Mag. Vol. 89, pp. 641–656

Holzapfel-Meyer, M. (1943). Affektive Grundlagen tierischen Verhaltens. Schweiz. Zs. Psychol. Vol. 2, pp. 19–42

Howard, E. (1920). Territory in Bird Life. London

Janet, P. (1935). Les débuts de l'intelligence. Paris

Jennings, H. S. (1906). Behaviour of the lower organisms. New York

Kalischer, O. (1907). Einige Bemerkungen über meine Dressurmethode. Zentralbl. Physiol. Vol. 21, pp. 585–586

Katz. D. (1927). Charakterologie und Tierpsychologie. Jahrb. Charakterol. Vol. 4, pp. 359–384

Katz, D. (1948). Mensch und Tier. Studien zur vergleichenden Psychologie. Zürich

Katz, I. (1949). Behavioral Interactions in a Herd of Barbary Sheep (Ammotragus lervia). Zoologica Vol. 34, pp. 9–18

Kellog, W. N. and Kellog, L. A. (1938). The Ape and the Child. New York and London

Kietz, G. (1952). Der Ausdrucksgehalt des menschlichen Ganges. Leipzig

Klatt, B. (1927). Entstehung der Haustiere. Handb. Vererbugswiss. Vol. 3. Berlin

Knoll, Fr. (1919). Gibt es eine Farbendressur der Insekten ? Die Naturwiss. Vol. 7

Knottnerus-Meyer, Th. (1924). Tiere im Zoo. Beobachtung eines Tierfreundes. Leipzig

Koch, W. (1951). Psychogene Beeinflussung des Geburtstermins bei Pferden. Zs. f. Tierpsychol. Vol. 8, pp. 441–443

Koehler, O. (1928). Untersuchungsmethoden der allgemeinen Reizphysiologie und der Verhaltensforschung an Tieren. Method. wiss. Biol. Vol. 2. Berlin

Kretschmer, E. (1930). Medizinische Psychologie, Leipzig

Krieg, H. (1929). Biologische Reisestudien in Südamerika. IX. Gürteltiere. Zs. Morphol. Ökol. Tiere. Vol. 14, pp. 166–190

Krieg, H. (1940). Als Zoologe in Steppen und Wäldern Patagoniens. Munich

Krumbiegel, I. (1947). Von Haustieren und ihrer Geschichte. Stuttgart

Kuckuk, E. (1936). Tierpsychologische Beobachtungen an zwei jungen Braunbären. Zs. vgl. Physiol. Vol. 24, pp. 14–41

Lack, D. (1946). The Life of the Robin. London

Leffler, O. H. (1914.) Zur Psychologie und Biologie des Axolotls. Abh. Ber. Mus. f. Natur- und Heimatkunde. Naturwiss. Verein Magdeburg. Vol. 3

Linsdale, J. M. (1946). The California Ground Squirrel. Berkeley and Los Angeles

Lorenz, K. (1935). Der Kumpan in der Umwelt des Vogels. J. Ornithol. Berlin. Vol. 83, pp. 137–413

Lorenz, K. (1937). Über die Bildung des Instinktbegriffes. Die Naturwiss. 25. Jahrg. pp. 289–331

Lorenz, K. (1940). Durch Domestikation verursachte Störungen arteigenen Verhaltens. Zs. angew. Psychol. Charakterkunde. Vol. 59, pp. 1–81

Lorenz, K. (1943). Die angeborenen Formen möglicher Erfahrung. Zs. f. Tierpsychol. Vol. 5, pp. 235–409

Maday, St. von (1912). Psychologie des Pferdes und der Dressur. Berlin

McDougall, W. (1930). Second report on a Lamarckian experiment. The Brit. J. Psychol. (Gen. Sect.), Vol. 20, pp. 201–218

Menzel, R. and Menzel, R. (1930). Die Verwertung der Riechfähigkeit des Hundes im Dienste der Menschheit. Berlin

Menzel, R. (1937). Welpe und Umwelt. Leipzig

Mertens, R. (1942). Über das Verhalten der Spaltenschildkröte, Malacochersus tornieri (Siebenrock). D. Zoolog. Garten N.F. Vol. 14, pp. 245–251

Mertens, R. (1946). Die Warn- und Drohreaktionen der Reptilien. Abh. Senckenberg. Naturf. Ges. 471. pp. 1–108

Merz, W. R. (1948). Normale und pathologische Physiologie der Laktation. Bibl. Gynaecol. Vol. 8

Minchin, A. K. (1937). Notes on the weaning of a young Koala (Phascolarctus cinereus). Rec. South. Austral. Mus. Vol. 16

Möhres, F. P. (1951). Die Wochenstuben der Fledermäuse. Die Umschau. Vol. 51, pp. 658–660

Morgan, C. L. (1894). Introduction to comparative psychology. London

Munn, N. L. (1933). An introduction to animal psychology. The Behaviour of the Rat. The Riverside Press, Cambridge

Nachtsheim, H. (1949). Vom Wildtier zum Haustier. Berlin and Hamburg

Neal, E. (1948). The Badger. London

Nissen, H. W. and Crawford, M. P. (1936). A preliminary study of foodsharing behaviour in young chimpanzees. J. comp. Psychol. Vol. 22

Pawlow, J. P. (1926). Die höchste Nerventätigkeit (das Verhalten) von Tieren. München

Peiper, A. (1949). Die Eigenart der kindlichen Hirntätigkeit. Leipzig

Peters, H. M. (1948). Grundfragen der Tierpsychologie. Stuttgart

Pfister-Ammende, M. (1949). Psychologische Erfahrungen mit sowjetrussischen Flüchtlingen in der Schweiz. Die Psychohygiene, 2. Reihe. Vol. 2

Pfister-Ammende, M. (1950). Das Problem der Entwurzelung. Schweiz. med. Wochenschrift, 80. Jahrg., No. 6

Pitman, Ch. R. S. (1938). A Guide to the Snakes of Uganda. Kampala, Uganda

Podkopaew, N. A. (1926). Die Methodik der Erforschung der bedingten Reflexe. München

Pope, C. H. (1937). Snakes alive and how they live. New York

Portmann, A. (1953). Das Tier als soziales Wesen. Zürich

Procter, J. B. (1924). Unrecorded Characters seen in Living Snakes. Proc. Zool. Soc. London. Vol. 2, pp. 1125–1129

Reynolds, H. W. (1950). Golden Hamsters. Stafford

Rothmann, M. and Teuber, E. (1915). Einzelausgabe aus der Anthropoidenstation auf Teneriffa. 1 Abh. Preuss. Akad. Wiss. Berlin. pp. 1–20

de Sanctis, Sante (1901). Die Träume (Übersetzung). Halle a.d.S.

Sarasin, P. (1915). Über tierische und menschliche Schnellrechner. Verh. Naturf. Ges. Basel. Vol. 26, pp. 68–95

Sarris, E. G. (1935). Die Umwelt des Hundes. Die Welt im Fortschritt. Vol. 3. Berlin

Schaffer, J. (1940). Die Hautdrüsenorgane der Säugetiere. Berlin and Vienna

Scheitlin, P. (1840). Versuch einer vollständigen Thierseelenkunde. Stuttgart and Tübingen

Schenkel, R. (1947). Ausdrucksstudien an Wölfen. Behaviour. Vol. 1, pp. 81–129

Schjelderup-Ebbe, Th. (1921). Beiträge zur Biologie und Sozial- und Individualpsychologie bei Gallus domesticus. Greifswald

Schjelderup-Ebbe, Th. (1922). Beiträge zur Sozialpsychologie des Haushuhns. Zs. Psychol. Vol. 88

Schmid, B. (1930). Aus der Welt des Tieres. Berlin

Schmidt, U. (1935). Das Muffelwild. Neudamm-Berlin

Schneider, K. M. (1930). Beobachtungen über die Pupillengestalt bei einigen lebenden Säugetieren. Neue Psychol. Studien. Hrsg. Felix Krueger. Vol. 6. München

Schneider, K. M. (1933). Zur Jugendentwicklung eines Eisbären. D. Zoolog. Garten N.F. Vol. 6

Schneider, K. M. (1944). Geburt eines roten Riesenkänguruhs (Macropus rufus Desm.). D. Zoolog. Garten N. F. Vol. 16

Shaw, Ch. E. (1948). The Male Combat " Dance " of Some Crotalid Snakes. Herpetologica. Vol. 4, pp. 137–145

Shaw, Ch. E. (1951). Male Combat in American Colubrid Snakes with Remarks on Combat in Other Colubrid and Elapid Snakes. Herpetologica. Vol. 7, pp. 149–168

Singh, J. A. L. and Zingg, R. M. (1942). Wolf-children and Feral Man. New York and London

Slijper, E. J. (1949). On some phenomena concerning pregnancy and parturition of the cetacea. Bijdragen tot de Dierkunde. Vol. 28, pp. 416–448

Small, W. S. (1900). An experimental study of the mental processes of the rat. Amer. J. Psychol. Vols. 11–12

Snoo, K. de (1942). Das Problem der Menschwerdung im Lichte der vergleichenden Geburtshilfe. Jena

Southern, H. N. (1948). Sexual and aggresive behaviour in the wild rabbit. Behaviour. Vol. 1, pp. 173–194

Spindler, M. and Bluhm, E. (1934). Kleine Beiträge zur Psychologie des Seelöwen (Eumetopias californianus). Zs. vgl. Physiol. Vol. 21, pp. 616–631

Steinbacher, G. (1938). Zum Problem der Haustierwerdung. Zs. f. Tierpsychol. Vol. 2, pp. 302–313

Steinbacher, G. (1941). Geburt und Kindheit eines Schimpansen. Zs. f. Tierpsychol. Vol. 4, pp. 188–203

Steiniger, F. (1936). Die Biologie der sogenannten " tierischen Hypnose ". Ergebn. Biol. Berlin. Vol. 13, pp. 348–451

Steiniger, F. (1938). Warnen und Tarnen im Tierreich. Berlin-Lichterfelde

Stemmler, C. (1935). Beitrag zur Fortpflanzungsbiologie europäischer Colubridae. D. Zoolog. Garten N.F. Vol. 8, pp. 38–41

Sutter, E. (1946). Das Abwehrverhalten nestjunger Wiedehopfe. Der Ornithol. Beob. Vol. 43, pp. 72–81

Szymanski, J. S. (1920). Aktivität und Ruhe bei Tieren und Menschen. Zs. allg. Physiol. Vol. 28, pp. 105–162

ten Cate, J. (1938). Zur Physiologie des Zentralnervensystems des Amphioxus (Branchiostoma lanceolatum). I. Die reflektorische Tätigkeit des Amphioxus. Arch. Neérl. Physiol. homme et animaux. Vol. 23, pp. 409–415

Throughton, E. (1948). Furred Animals of Australia. Sydney-London

Tinbergen, N. (1940). Die Übersprungbewegung. Zs. f. Tierpsychol. Vol. 4, pp. 1–40

Tinbergen, N. (1951). The Study of Instinct. Oxford

Tinbergen, N. (1953). Social Behaviour in Animals. London

Uexküll, J. von (1940). Bedeutungslehre. Bios. Vol. 10. Leipzig

Urbain, A., Déchambre, E. and Rode, P. (1941). Observations faites sur un jeune Orang-Utan né à la ménagerie du Jardin des Plantes. Mammalia. pp. 82–85

Völgyesi, Franz (1938). Menschen- und Tierhypnose. Zürich-Leipzig

Washburn, M. (1930). The Animal Mind. New York

Weidmann, U. (1951). Über den systematischen Wert von Balzhandlungen bei Drosophila. Rev. Suisse Zool. Vol. 54, No. 28, pp. 502–511

Witmer, L. (1909). A monkey with a mind. Psychol. Clinic Philadelphia. Vol. 3, pp. 179–205

Wolff, G. (1914). Die denkenden Tiere von Elberfeld und Mannheim. Süddeut. Monatsschr. pp. 456–467

Wörner, R. (1940). Theoretische und experimentelle Beiträge zum Ausdrucksproblem. Zs. angew. Psychol. und Charakterkunde. Vol. 59, pp. 257–318

Yerkes, R. M. and Yerkes, A. (1934). The Great Apes. New Haven

Yerkes, R. M. (1948). Chimpanzees, a laboratory colony. New Haven.

Ziegler, H. E. (1920). Der Begriff des Instinktes einst und jetzt. Eine Studie über die Geschichte und die Grundlagen der Tierpsychologie. Jena

Zschokke, F. (1916). Der Schlaf der Tiere. Basle.

INDEX

A CATALOGUE OF SELECTED DOVER BOOKS
IN ALL FIELDS OF INTEREST

THE DEVIL'S DICTIONARY, Ambrose Bierce. Barbed, bitter, brilliant witticisms in the form of a dictionary. Best, most ferocious satire America has produced. 145pp. 20487-1 Pa. $1.75

ABSOLUTELY MAD INVENTIONS, A.E. Brown, H.A. Jeffcott. Hilarious, useless, or merely absurd inventions all granted patents by the U.S. Patent Office. Edible tie pin, mechanical hat tipper, etc. 57 illustrations. 125pp. 22596-8 Pa. $1.50

AMERICAN WILD FLOWERS COLORING BOOK, Paul Kennedy. Planned coverage of 48 most important wildflowers, from Rickett's collection; instructive as well as entertaining. Color versions on covers. 48pp. 8¼ x 11. 20095-7 Pa. $1.50

BIRDS OF AMERICA COLORING BOOK, John James Audubon. Rendered for coloring by Paul Kennedy. 46 of Audubon's noted illustrations: red-winged blackbird, cardinal, purple finch, towhee, etc. Original plates reproduced in full color on the covers. 48pp. 8¼ x 11. 23049-X Pa. $1.35

NORTH AMERICAN INDIAN DESIGN COLORING BOOK, Paul Kennedy. The finest examples from Indian masks, beadwork, pottery, etc. — selected and redrawn for coloring (with identifications) by well-known illustrator Paul Kennedy. 48pp. 8¼ x 11. 21125-8 Pa. $1.35

UNIFORMS OF THE AMERICAN REVOLUTION COLORING BOOK, Peter Copeland. 31 lively drawings reproduce whole panorama of military attire; each uniform has complete instructions for accurate coloring. (Not in the Pictorial Archives Series). 64pp. 8¼ x 11. 21850-3 Pa. $1.50

THE WONDERFUL WIZARD OF OZ COLORING BOOK, L. Frank Baum. Color the Yellow Brick Road and much more in 61 drawings adapted from W.W. Denslow's originals, accompanied by abridged version of text. Dorothy, Toto, Oz and the Emerald City. 61 illustrations. 64pp. 8¼ x 11. 20452-9 Pa. $1.50

CUT AND COLOR PAPER MASKS, Michael Grater. Clowns, animals, funny faces . . . simply color them in, cut them out, and put them together, and you have 9 paper masks to play with and enjoy. Complete instructions. Assembled masks shown in full color on the covers. 32pp. 8¼ x 11. 23171-2 Pa. $1.50

STAINED GLASS CHRISTMAS ORNAMENT COLORING BOOK, Carol Belanger Grafton. Brighten your Christmas season with over 100 Christmas ornaments done in a stained glass effect on translucent paper. Color them in and then hang at windows, from lights, anywhere. 32pp. 8¼ x 11. 20707-2 Pa. $1.75

A CATALOGUE OF SELECTED DOVER BOOKS
IN ALL FIELDS OF INTEREST

THE ART DECO STYLE, ed. by Theodore Menten. Furniture, jewelry, metalwork, ceramics, fabrics, lighting fixtures, interior decors, exteriors, graphics from pure French sources. Best sampling around. Over 400 photographs. 183pp. 8⅜ x 11¼.
22824-X Pa. $4.00

THE GENTLEMAN AND CABINET MAKER'S DIRECTOR, Thomas Chippendale. Full reprint, 1762 style book, most influential of all time; chairs, tables, sofas, mirrors, cabinets, etc. 200 plates, plus 24 photographs of surviving pieces. 249pp. 9⅞ x 12¾.
21601-2 Pa. $5.00

PINE FURNITURE OF EARLY NEW ENGLAND, Russell H. Kettell. Basic book. Thorough historical text, plus 200 illustrations of boxes, highboys, candlesticks, desks, etc. 477pp. 7⅞ x 10¾.
20145-7 Clothbd. $12.50

ORIENTAL RUGS, ANTIQUE AND MODERN, Walter A. Hawley. Persia, Turkey, Caucasus, Central Asia, China, other traditions. Best general survey of all aspects: styles and periods, manufacture, uses, symbols and their interpretation, and identification. 96 illustrations, 11 in color. 320pp. 6⅛ x 9¼.
22366-3 Pa. $5.00

DECORATIVE ANTIQUE IRONWORK, Henry R. d'Allemagne. Photographs of 4500 iron artifacts from world's finest collection, Rouen. Hinges, locks, candelabra, weapons, lighting devices, clocks, tools, from Roman times to mid-19th century. Nothing else comparable to it. 420pp. 9 x 12.
22082-6 Pa. $8.50

THE COMPLETE BOOK OF DOLL MAKING AND COLLECTING, Catherine Christopher. Instructions, patterns for dozens of dolls, from rag doll on up to elaborate, historically accurate figures. Mould faces, sew clothing, make doll houses, etc. Also collecting information. Many illustrations. 288pp. 6 x 9. 22066-4 Pa. $3.00

ANTIQUE PAPER DOLLS: 1915-1920, edited by Arnold Arnold. 7 antique cut-out dolls and 24 costumes from 1915-1920, selected by Arnold Arnold from his collection of rare children's books and entertainments, all in full color. 32pp. 9¼ x 12¼.
23176-3 Pa. $2.00

ANTIQUE PAPER DOLLS: THE EDWARDIAN ERA, Epinal. Full-color reproductions of two historic series of paper dolls that show clothing styles in 1908 and at the beginning of the First World War. 8 two-sided, stand-up dolls and 32 complete, two-sided costumes. Full instructions for assembling included. 32pp. 9¼ x 12¼.
23175-5 Pa. $2.00

A HISTORY OF COSTUME, Carl Köhler, Emma von Sichardt. Egypt, Babylon, Greece up through 19th century Europe; based on surviving pieces, art works, etc. Full text and 595 illustrations, including many clear, measured patterns for reproducing historic costume. Practical. 464pp. 21030-8 Pa. $4.00

EARLY AMERICAN LOCOMOTIVES, John H. White, Jr. Finest locomotive engravings from late 19th century: historical (1804-1874), main-line (after 1870), special, foreign, etc. 147 plates. 200pp. 11⅜ x 8¼. 22772-3 Pa. $3.50

VICTORIAN HOUSES: A TREASURY OF LESSER-KNOWN EXAMPLES, Edmund Gillon and Clay Lancaster. 116 photographs, excellent commentary illustrate distinct characteristics, many borrowings of local Victorian architecture. Octagonal houses, Americanized chalets, grand country estates, small cottages, etc. Rich heritage often overlooked. 116 plates. 11⅜ x 10. 22966-1 Pa. $4.00

STICKS AND STONES, Lewis Mumford. Great classic of American cultural history; architecture from medieval-inspired earliest forms to 20th century; evolution of structure and style, influence of environment. 21 illustrations. 113pp.
20202-X Pa. $2.00

ON THE LAWS OF JAPANESE PAINTING, Henry P. Bowie. Best substitute for training with genius Oriental master, based on years of study in Kano school. Philosophy, brushes, inks, style, etc. 66 illustrations. 117pp. 6⅛ x 9¼. 20030-2 Pa. $4.00

A HANDBOOK OF ANATOMY FOR ART STUDENTS, Arthur Thomson. Virtually exhaustive. Skeletal structure, muscles, heads, special features. Full text, anatomical figures, undraped photos. Male and female. 337 illustrations. 459pp.
21163-0 Pa. $5.00

AN ATLAS OF ANATOMY FOR ARTISTS, Fritz Schider. Finest text, working book. Full text, plus anatomical illustrations; plates by great artists showing anatomy. 593 illustrations. 192pp. 7⅞ x 10¾. 20241-0 Clothbd. $6.95

THE HUMAN FIGURE IN MOTION, Eadweard Muybridge. More than 4500 stopped-action photos, in action series, showing undraped men, women, children jumping, lying down, throwing, sitting, wrestling, carrying, etc. "Unparalleled dictionary for artists," American Artist. Taken by great 19th century photographer. 390pp. 7⅞ x 10⅝. 20204-6 Clothbd. $12.50

AN ATLAS OF ANIMAL ANATOMY FOR ARTISTS, W. Ellenberger et al. Horses, dogs, cats, lions, cattle, deer, etc. Muscles, skeleton, surface features. The basic work. Enlarged edition. 288 illustrations. 151pp. 9⅜ x 12¼. 20082-5 Pa. $4.00

LETTER FORMS: 110 COMPLETE ALPHABETS, Frederick Lambert. 110 sets of capital letters; 16 lower case alphabets; 70 sets of numbers and other symbols. Edited and expanded by Theodore Menten. 110pp. 8⅛ x 11. 22872-X Pa. $2.50

THE METHODS OF CONSTRUCTION OF CELTIC ART, George Bain. Simple geometric techniques for making wonderful Celtic interlacements, spirals, Kells-type initials, animals, humans, etc. Unique for artists, craftsmen. Over 500 illustrations. 160pp. 9 x 12. USO 22923-8 Pa. $4.00

SCULPTURE, PRINCIPLES AND PRACTICE, Louis Slobodkin. Step by step approach to clay, plaster, metals, stone; classical and modern. 253 drawings, photos. 255pp. 8⅛ x 11. 22960-2 Pa. $4.50

THE ART OF ETCHING, E.S. Lumsden. Clear, detailed instructions for etching, drypoint, softground, aquatint; from 1st sketch to print. Very detailed, thorough. 200 illustrations. 376pp. 20049-3 Pa. $3.50

JEWISH GREETING CARDS, Ed Sibbett, Jr. 16 cards to cut and color. Three say "Happy Chanukah," one "Happy New Year," others have no message, show stars of David, Torahs, wine cups, other traditional themes. 16 envelopes. 8¼ x 11.
23225-5 Pa. $2.00

AUBREY BEARDSLEY GREETING CARD BOOK, Aubrey Beardsley. Edited by Theodore Menten. 16 elegant yet inexpensive greeting cards let you combine your own sentiments with subtle Art Nouveau lines. 16 different Aubrey Beardsley designs that you can color or not, as you wish. 16 envelopes. 64pp. 8¼ x 11.
23173-9 Pa. $2.00

RECREATIONS IN THE THEORY OF NUMBERS, Albert Beiler. Number theory, an inexhaustible source of puzzles, recreations, for beginners and advanced. Divisors, perfect numbers. scales of notation, etc. 349pp.
21096-0 Pa. $2.50

AMUSEMENTS IN MATHEMATICS, Henry E. Dudeney. One of largest puzzle collections, based on algebra, arithmetic, permutations, probability, plane figure dissection, properties of numbers, by one of world's foremost puzzlists. Solutions. 450 illustrations. 258pp.
20473-1 Pa. $2.75

MATHEMATICS, MAGIC AND MYSTERY, Martin Gardner. Puzzle editor for Scientific American explains math behind: card tricks, stage mind reading, coin and match tricks, counting out games, geometric dissections. Probability, sets, theory of numbers, clearly explained. Plus more than 400 tricks, guaranteed to work. 135 illustrations. 176pp.
20335-2 Pa. $2.00

BEST MATHEMATICAL PUZZLES OF SAM LOYD, edited by Martin Gardner. Bizarre, original, whimsical puzzles by America's greatest puzzler. From fabulously rare Cyclopedia, including famous 14-15 puzzles, the Horse of a Different Color, 115 more. Elementary math. 150 illustrations. 167pp.
20498-7 Pa. $2.00

MATHEMATICAL PUZZLES FOR BEGINNERS AND ENTHUSIASTS, Geoffrey Mott-Smith. 189 puzzles from easy to difficult involving arithmetic, logic, algebra, properties of digits, probability. Explanation of math behind puzzles. 135 illustrations. 248pp.
20198-8 Pa.$2.75

BIG BOOK OF MAZES AND LABYRINTHS, Walter Shepherd. Classical, solid, and ripple mazes; short path and avoidance labyrinths; more — 50 mazes and labyrinths in all. 12 other figures. Full solutions. 112pp. 8⅛ x 11.
22951-3 Pa. $2.00

COIN GAMES AND PUZZLES, Maxey Brooke. 60 puzzles, games and stunts — from Japan, Korea, Africa and the ancient world, by Dudeney and the other great puzzlers, as well as Maxey Brooke's own creations. Full solutions. 67 illustrations. 94pp.
22893-2 Pa. $1.25

HAND SHADOWS TO BE THROWN UPON THE WALL, Henry Bursill. Wonderful Victorian novelty tells how to make flying birds, dog, goose, deer, and 14 others. 32pp. 6½ x 9¼.
21779-5 Pa. $1.25

DECORATIVE ALPHABETS AND INITIALS, edited by Alexander Nesbitt. 91 complete alphabets (medieval to modern), 3924 decorative initials, including Victorian novelty and Art Nouveau. 192pp. 7¾ x 10¾. 20544-4 Pa. $3.50

CALLIGRAPHY, Arthur Baker. Over 100 original alphabets from the hand of our greatest living calligrapher: simple, bold, fine-line, richly ornamented, etc. — all strikingly original and different, a fusion of many influences and styles. 155pp. 11⅜ x 8¼. 22895-9 Pa. $4.00

MONOGRAMS AND ALPHABETIC DEVICES, edited by Hayward and Blanche Cirker. Over 2500 combinations, names, crests in very varied styles: script engraving, ornate Victorian, simple Roman, and many others. 226pp. 8⅛ x 11. 22330-2 Pa. $5.00

THE BOOK OF SIGNS, Rudolf Koch. Famed German type designer renders 493 symbols: religious, alchemical, imperial, runes, property marks, etc. Timeless. 104pp. 6⅛ x 9¼. 20162-7 Pa. $1.50

200 DECORATIVE TITLE PAGES, edited by Alexander Nesbitt. 1478 to late 1920's. Baskerville, Dürer, Beardsley, W. Morris, Pyle, many others in most varied techniques. For posters, programs, other uses. 222pp. 8⅜ x 11¼. 21264-5 Pa. $3.50

DICTIONARY OF AMERICAN PORTRAITS, edited by Hayward and Blanche Cirker. 4000 important Americans, earliest times to 1905, mostly in clear line. Politicians, writers, soldiers, scientists, inventors, industrialists, Indians, Blacks, women, outlaws, etc. Identificatory information. 756pp. 9¼ x 12¾. 21823-6 Clothbd. $30.00

ART FORMS IN NATURE, Ernst Haeckel. Multitude of strangely beautiful natural forms: Radiolaria, Foraminifera, jellyfishes, fungi, turtles, bats, etc. All 100 plates of the 19th century evolutionist's Kunstformen der Natur (1904). 100pp. 9⅜ x 12¼. 22987-4 Pa. $4.00

DECOUPAGE: THE BIG PICTURE SOURCEBOOK, Eleanor Rawlings. Make hundreds of beautiful objects, over 550 florals, animals, letters, shells, period costumes, frames, etc. selected by foremost practitioner. Printed on one side of page. 8 color plates. Instructions. 176pp. 9³/₁₆ x 12¼. 23182-8 Pa. $5.00

AMERICAN FOLK DECORATION, Jean Lipman, Eve Meulendyke. Thorough coverage of all aspects of wood, tin, leather, paper, cloth decoration — scapes, humans, trees, flowers, geometrics — and how to make them. Full instructions. 233 illustrations, 5 in color. 163pp. 8⅜ x 11¼. 22217-9 Pa. $3.95

WHITTLING AND WOODCARVING, E.J. Tangerman. Best book on market; clear, full. If you can cut a potato, you can carve toys, puzzles, chains, caricatures, masks, patterns, frames, decorate surfaces, etc. Also covers serious wood sculpture. Over 200 photos. 293pp. 20965-2 Pa. $2.50

THE JOURNAL OF HENRY D. THOREAU, edited by Bradford Torrey, F.H. Allen. Complete reprinting of 14 volumes, 1837-1861, over two million words; the source-books for Walden, etc. Definitive. All original sketches, plus 75 photographs. Introduction by Walter Harding. Total of 1804pp. 8½ x 12¼.
20312-3, 20313-1 Clothbd., Two vol. set $50.00

MASTERS OF THE DRAMA, John Gassner. Most comprehensive history of the drama, every tradition from Greeks to modern Europe and America, including Orient. Covers 800 dramatists, 2000 plays; biography, plot summaries, criticism, theatre history, etc. 77 illustrations. 890pp. 20100-7 Clothbd. $10.00

GHOST AND HORROR STORIES OF AMBROSE BIERCE, Ambrose Bierce. 23 modern horror stories: The Eyes of the Panther, The Damned Thing, etc., plus the dream-essay Visions of the Night. Edited by E.F. Bleiler. 199pp. 20767-6 Pa. $2.00

BEST GHOST STORIES, Algernon Blackwood. 13 great stories by foremost British 20th century supernaturalist. The Willows, The Wendigo, Ancient Sorceries, others. Edited by E.F. Bleiler. 366pp. USO 22977-7 Pa. $3.00

THE BEST TALES OF HOFFMANN, E.T.A. Hoffmann. 10 of Hoffmann's most important stories, in modern re-editings of standard translations: Nutcracker and the King of Mice, The Golden Flowerpot, etc. 7 illustrations by Hoffmann. Edited by E.F. Bleiler. 458pp. 21793-0 Pa. $3.95

BEST GHOST STORIES OF J.S. LEFANU, J. Sheridan LeFanu. 16 stories by greatest Victorian master: Green Tea, Carmilla, Haunted Baronet, The Familiar, etc. Mostly unavailable elsewhere. Edited by E.F. Bleiler. 8 illustrations. 467pp.
20415-4 Pa. $4.00

SUPERNATURAL HORROR IN LITERATURE, H.P. Lovecraft. Great modern American supernaturalist brilliantly surveys history of genre to 1930's, summarizing, evaluating scores of books. Necessary for every student, lover of form. Introduction by E.F. Bleiler. 111pp. 20105-8 Pa. $1.50

THREE GOTHIC NOVELS, ed. by E.F. Bleiler. Full texts Castle of Otranto, Walpole; Vathek, Beckford; The Vampyre, Polidori; Fragment of a Novel, Lord Byron. 331pp. 21232-7 Pa. $3.00

SEVEN SCIENCE FICTION NOVELS, H.G. Wells. Full novels. First Men in the Moon, Island of Dr. Moreau, War of the Worlds, Food of the Gods, Invisible Man, Time Machine, In the Days of the Comet. A basic science-fiction library. 1015pp.
USO 20264-X Clothbd. $6.00

LADY AUDLEY'S SECRET, Mary E. Braddon. Great Victorian mystery classic, beautifully plotted, suspenseful; praised by Thackeray, Boucher, Starrett, others. What happened to beautiful, vicious Lady Audley's husband? Introduction by Norman Donaldson. 286pp. 23011-2 Pa. $3.00

SLEEPING BEAUTY, illustrated by Arthur Rackham. Perhaps the fullest, most delightful version ever, told by C.S. Evans. Rackham's best work. 49 illustrations. 110pp. 7⅞ x 10¾. 22756-1 Pa. $2.00

THE WONDERFUL WIZARD OF OZ, L. Frank Baum. Facsimile in full color of America's finest children's classic. Introduction by Martin Gardner. 143 illustrations by W.W. Denslow. 267pp. 20691-2 Pa. $2.50

GOOPS AND HOW TO BE THEM, Gelett Burgess. Classic tongue-in-cheek masquerading as etiquette book. 87 verses, 170 cartoons as Goops demonstrate virtues of table manners, neatness, courtesy, more. 88pp. 6½ x 9¼. 22233-0 Pa. $1.50

THE BROWNIES, THEIR BOOK, Palmer Cox. Small as mice, cunning as foxes, exuberant, mischievous, Brownies go to zoo, toy shop, seashore, circus, more. 24 verse adventures. 266 illustrations. 144pp. 6⅝ x 9¼. 21265-3 Pa. $1.75

BILLY WHISKERS: THE AUTOBIOGRAPHY OF A GOAT, Frances Trego Montgomery. Escapades of that rambunctious goat. Favorite from turn of the century America. 24 illustrations. 259pp. 22345-0 Pa. $2.75

THE ROCKET BOOK, Peter Newell. Fritz, janitor's kid, sets off rocket in basement of apartment house; an ingenious hole punched through every page traces course of rocket. 22 duotone drawings, verses. 48pp. 6⅞ x 8⅜. 22044-3 Pa. $1.50

PECK'S BAD BOY AND HIS PA, George W. Peck. Complete double-volume of great American childhood classic. Hennery's ingenious pranks against outraged pomposity of pa and the grocery man. 97 illustrations. Introduction by E.F. Bleiler. 347pp. 20497-9 Pa. $2.50

THE TALE OF PETER RABBIT, Beatrix Potter. The inimitable Peter's terrifying adventure in Mr. McGregor's garden, with all 27 wonderful, full-color Potter illustrations. 55pp. 4¼ x 5½. USO 22827-4 Pa. $1.00

THE TALE OF MRS. TIGGY-WINKLE, Beatrix Potter. Your child will love this story about a very special hedgehog and all 27 wonderful, full-color Potter illustrations. 57pp. 4¼ x 5½. USO 20546-0 Pa. $1.00

THE TALE OF BENJAMIN BUNNY, Beatrix Potter. Peter Rabbit's cousin coaxes him back into Mr. McGregor's garden for a whole new set of adventures. A favorite with children. All 27 full-color illustrations. 59pp. 4¼ x 5½. USO 21102-9 Pa. $1.00

THE MERRY ADVENTURES OF ROBIN HOOD, Howard Pyle. Facsimile of original (1883) edition, finest modern version of English outlaw's adventures. 23 illustrations by Pyle. 296pp. 6½ x 9¼. 22043-5 Pa. $2.75

TWO LITTLE SAVAGES, Ernest Thompson Seton. Adventures of two boys who lived as Indians; explaining Indian ways, woodlore, pioneer methods. 293 illustrations. 286pp. 20985-7 Pa. $3.00

CATALOGUE OF DOVER BOOKS

THE MAGIC MOVING PICTURE BOOK, Bliss, Sands & Co. The pictures in this book move! Volcanoes erupt, a house burns, a serpentine dancer wiggles her way through a number. By using a specially ruled acetate screen provided, you can obtain these and 15 other startling effects. Originally "The Motograph Moving Picture Book." 32pp. 8¼ x 11.   23224-7 Pa. $1.75

STRING FIGURES AND HOW TO MAKE THEM, Caroline F. Jayne. Fullest, clearest instructions on string figures from around world: Eskimo, Navajo, Lapp, Europe, more. Cats cradle, moving spear, lightning, stars. Introduction by A.C. Haddon. 950 illustrations. 407pp.   20152-X Pa. $3.00

PAPER FOLDING FOR BEGINNERS, William D. Murray and Francis J. Rigney. Clearest book on market for making origami sail boats, roosters, frogs that move legs, cups, bonbon boxes. 40 projects. More than 275 illustrations. Photographs. 94pp.
                20713-7 Pa. $1.25

INDIAN SIGN LANGUAGE, William Tomkins. Over 525 signs developed by Sioux, Blackfoot, Cheyenne, Arapahoe and other tribes. Written instructions and diagrams: how to make words, construct sentences. Also 290 pictographs of Sioux and Ojibway tribes. 111pp. 6⅛ x 9¼.   22029-X Pa. $1.50

BOOMERANGS: HOW TO MAKE AND THROW THEM, Bernard S. Mason. Easy to make and throw, dozens of designs: cross-stick, pinwheel, boomabird, tumblestick, Australian curved stick boomerang. Complete throwing instructions. All safe. 99pp.   23028-7 Pa. $1.50

25 KITES THAT FLY, Leslie Hunt. Full, easy to follow instructions for kites made from inexpensive materials. Many novelties. Reeling, raising, designing your own. 70 illustrations. 110pp.   22550-X Pa. $1.25

TRICKS AND GAMES ON THE POOL TABLE, Fred Herrmann. 79 tricks and games, some solitaires, some for 2 or more players, some competitive; mystifying shots and throws, unusual carom, tricks involving cork, coins, a hat, more. 77 figures. 95pp.   21814-7 Pa. $1.25

WOODCRAFT AND CAMPING, Bernard S. Mason. How to make a quick emergency shelter, select woods that will burn immediately, make do with limited supplies, etc. Also making many things out of wood, rawhide, bark, at camp. Formerly titled Woodcraft. 295 illustrations. 580pp.   21951-8 Pa. $4.00

AN INTRODUCTION TO CHESS MOVES AND TACTICS SIMPLY EXPLAINED, Leonard Barden. Informal intermediate introduction: reasons for moves, tactics, openings, traps, positional play, endgame. Isolates patterns. 102pp. USO 21210-6 Pa. $1.35

LASKER'S MANUAL OF CHESS, Dr. Emanuel Lasker. Great world champion offers very thorough coverage of all aspects of chess. Combinations, position play, openings, endgame, aesthetics of chess, philosophy of struggle, much more. Filled with analyzed games. 390pp.   20640-8 Pa. $3.50

HOW TO SOLVE CHESS PROBLEMS, Kenneth S. Howard. Practical suggestions on problem solving for very beginners. 58 two-move problems, 46 3-movers, 8 4-movers for practice, plus hints. 171pp. 20748-X Pa. $2.00

A GUIDE TO FAIRY CHESS, Anthony Dickins. 3-D chess, 4-D chess, chess on a cylindrical board, reflecting pieces that bounce off edges, cooperative chess, retrograde chess, maximummers, much more. Most based on work of great Dawson. Full handbook, 100 problems. 66pp. 7⅞ x 10¾. 22687-5 Pa. $2.00

WIN AT BACKGAMMON, Millard Hopper. Best opening moves, running game, blocking game, back game, tables of odds, etc. Hopper makes the game clear enough for anyone to play, and win. 43 diagrams. 111pp. 22894-0 Pa. $1.50

BIDDING A BRIDGE HAND, Terence Reese. Master player "thinks out loud" the binding of 75 hands that defy point count systems. Organized by bidding problem—no-fit situations, overbidding, underbidding, cueing your defense, etc. 254pp. EBE 22830-4 Pa. $2.50

THE PRECISION BIDDING SYSTEM IN BRIDGE, C.C. Wei, edited by Alan Truscott. Inventor of precision bidding presents average hands and hands from actual play, including games from 1969 Bermuda Bowl where system emerged. 114 exercises. 116pp. 21171-1 Pa. $1.75

LEARN MAGIC, Henry Hay. 20 simple, easy-to-follow lessons on magic for the new magician: illusions, card tricks, silks, sleights of hand, coin manipulations, escapes, and more —all with a minimum amount of equipment. Final chapter explains the great stage illusions. 92 illustrations. 285pp. 21238-6 Pa. $2.95

THE NEW MAGICIAN'S MANUAL, Walter B. Gibson. Step-by-step instructions and clear illustrations guide the novice in mastering 36 tricks; much equipment supplied on 16 pages of cut-out materials. 36 additional tricks. 64 illustrations. 159pp. 6⅝ x 10. 23113-5 Pa. $3.00

PROFESSIONAL MAGIC FOR AMATEURS, Walter B. Gibson. 50 easy, effective tricks used by professionals —cards, string, tumblers, handkerchiefs, mental magic, etc. 63 illustrations. 223pp. 23012-0 Pa. $2.50

CARD MANIPULATIONS, Jean Hugard. Very rich collection of manipulations; has taught thousands of fine magicians tricks that are really workable, eye-catching. Easily followed, serious work. Over 200 illustrations. 163pp. 20539-8 Pa. $2.00

ABBOTT'S ENCYCLOPEDIA OF ROPE TRICKS FOR MAGICIANS, Stewart James. Complete reference book for amateur and professional magicians containing more than 150 tricks involving knots, penetrations, cut and restored rope, etc. 510 illustrations. Reprint of 3rd edition. 400pp. 23206-9 Pa. $3.50

THE SECRETS OF HOUDINI, J.C. Cannell. Classic study of Houdini's incredible magic, exposing closely-kept professional secrets and revealing, in general terms, the whole art of stage magic. 67 illustrations. 279pp. 22913-0 Pa. $2.50

DRIED FLOWERS, Sarah Whitlock and Martha Rankin. Concise, clear, practical guide to dehydration, glycerinizing, pressing plant material, and more. Covers use of silica gel. 12 drawings. Originally titled "New Techniques with Dried Flowers." 32pp. 21802-3 Pa. $1.00

ABC OF POULTRY RAISING, J.H. Florea. Poultry expert, editor tells how to raise chickens on home or small business basis. Breeds, feeding, housing, laying, etc. Very concrete, practical. 50 illustrations. 256pp. 23201-8 Pa. $3.00

HOW INDIANS USE WILD PLANTS FOR FOOD, MEDICINE & CRAFTS, Frances Densmore. Smithsonian, Bureau of American Ethnology report presents wealth of material on nearly 200 plants used by Chippewas of Minnesota and Wisconsin. 33 plates plus 122pp. of text. 6$\frac{1}{8}$ x 9$\frac{1}{4}$. 23019-8 Pa. $2.50

THE HERBAL OR GENERAL HISTORY OF PLANTS, John Gerard. The 1633 edition revised and enlarged by Thomas Johnson. Containing almost 2850 plant descriptions and 2705 superb illustrations, Gerard's Herbal is a monumental work, the book all modern English herbals are derived from, and the one herbal every serious enthusiast should have in its entirety. Original editions are worth perhaps $750. 1678pp. 8$\frac{1}{2}$ x 12$\frac{1}{4}$. 23147-X Clothbd. $50.00

A MODERN HERBAL, Margaret Grieve. Much the fullest, most exact, most useful compilation of herbal material. Gigantic alphabetical encyclopedia, from aconite to zedoary, gives botanical information, medical properties, folklore, economic uses, and much else. Indispensable to serious reader. 161 illustrations. 888pp. 6$\frac{1}{2}$ x 9$\frac{1}{4}$. USO 22798-7, 22799-5 Pa., Two vol. set $10.00

HOW TO KNOW THE FERNS, Frances T. Parsons. Delightful classic. Identification, fern lore, for Eastern and Central U.S.A. Has introduced thousands to interesting life form. 99 illustrations. 215pp. 20740-4 Pa. $2.50

THE MUSHROOM HANDBOOK, Louis C.C. Krieger. Still the best popular handbook. Full descriptions of 259 species, extremely thorough text, habitats, luminescence, poisons, folklore, etc. 32 color plates; 126 other illustrations. 560pp. 21861-9 Pa. $4.50

HOW TO KNOW THE WILD FRUITS, Maude G. Peterson. Classic guide covers nearly 200 trees, shrubs, smaller plants of the U.S. arranged by color of fruit and then by family. Full text provides names, descriptions, edibility, uses. 80 illustrations. 400pp. 22943-2 Pa. $3.00

COMMON WEEDS OF THE UNITED STATES, U.S. Department of Agriculture. Covers 220 important weeds with illustration, maps, botanical information, plant lore for each. Over 225 illustrations. 463pp. 6$\frac{1}{8}$ x 9$\frac{1}{4}$. 20504-5 Pa. $4.50

HOW TO KNOW THE WILD FLOWERS, Mrs. William S. Dana. Still best popular book for East and Central USA. Over 500 plants easily identified, with plant lore; arranged according to color and flowering time. 174 plates. 459pp. 20332-8 Pa. $3.50

AUSTRIAN COOKING AND BAKING, Gretel Beer. Authentic thick soups, wiener schnitzel, veal goulash, more, plus dumplings, puff pastries, nut cakes, sacher tortes, other great Austrian desserts. 224pp. USO 23220-4 Pa. $2.50

CHEESES OF THE WORLD, U.S.D.A. Dictionary of cheeses containing descriptions of over 400 varieties of cheese from common Cheddar to exotic Surati. Up to two pages are given to important cheeses like Camembert, Cottage, Edam, etc. 151pp. 22831-2 Pa. $1.50

TRITTON'S GUIDE TO BETTER WINE AND BEER MAKING FOR BEGINNERS, S.M. Tritton. All you need to know to make family-sized quantities of over 100 types of grape, fruit, herb, vegetable wines; plus beers, mead, cider, more. 11 illustrations. 157pp. USO 22528-3 Pa. $2.00

DECORATIVE LABELS FOR HOME CANNING, PRESERVING, AND OTHER HOUSEHOLD AND GIFT USES, Theodore Menten. 128 gummed, perforated labels, beautifully printed in 2 colors. 12 versions in traditional, Art Nouveau, Art Deco styles. Adhere to metal, glass, wood, most plastics. 24pp. 8¼ x 11. 23219-0 Pa. $2.00

FIVE ACRES AND INDEPENDENCE, Maurice G. Kains. Great back-to-the-land classic explains basics of self-sufficient farming: economics, plants, crops, animals, orchards, soils, land selection, host of other necessary things. Do not confuse with skimpy faddist literature; Kains was one of America's greatest agriculturalists. 95 illustrations. 397pp. 20974-1 Pa. $2.95

GROWING VEGETABLES IN THE HOME GARDEN, U.S. Dept. of Agriculture. Basic information on site, soil conditions, selection of vegetables, planting, cultivation, gathering. Up-to-date, concise, authoritative. Covers 60 vegetables. 30 illustrations. 123pp. 23167-4 Pa. $1.35

FRUITS FOR THE HOME GARDEN, Dr. U.P. Hedrick. A chapter covering each type of garden fruit, advice on plant care, soils, grafting, pruning, sprays, transplanting, and much more! Very full. 53 illustrations. 175pp. 22944-0 Pa. $2.50

GARDENING ON SANDY SOIL IN NORTH TEMPERATE AREAS, Christine Kelway. Is your soil too light, too sandy? Improve your soil, select plants that survive under such conditions. Both vegetables and flowers. 42 photos. 148pp.
USO 23199-2 Pa. $2.50

THE FRAGRANT GARDEN: A BOOK ABOUT SWEET SCENTED FLOWERS AND LEAVES, Louise Beebe Wilder. Fullest, best book on growing plants for their fragrances. Descriptions of hundreds of plants, both well-known and overlooked. 407pp.
23071-6 Pa. $3.50

EASY GARDENING WITH DROUGHT-RESISTANT PLANTS, Arno and Irene Nehrling. Authoritative guide to gardening with plants that require a minimum of water: seashore, desert, and rock gardens; house plants; annuals and perennials; much more. 190 illustrations. 320pp. 23230-1 Pa. $3.50

THE STYLE OF PALESTRINA AND THE DISSONANCE, Knud Jeppesen. Standard analysis of rhythm, line, harmony, accented and unaccented dissonances. Also pre-Palestrina dissonances. 306pp. 22386-8 Pa. $3.00

DOVER OPERA GUIDE AND LIBRETTO SERIES prepared by Ellen H. Bleiler. Each volume contains everything needed for background, complete enjoyment: complete libretto, new English translation with all repeats, biography of composer and librettist, early performance history, musical lore, much else. All volumes lavishly illustrated with performance photos, portraits, similar material. Do not confuse with skimpy performance booklets.

CARMEN, Georges Bizet. 66 illustrations. 222pp. 22111-3 Pa. $2.00
DON GIOVANNI, Wolfgang A. Mozart. 92 illustrations. 209pp. 21134-7 Pa. $2.50
LA BOHÈME, Giacomo Puccini. 73 illustrations. 124pp. USO 20404-9 Pa. $1.75
ÄIDA, Giuseppe Verdi. 76 illustrations. 181pp. 20405-7 Pa. $2.25
LUCIA DI LAMMERMOOR, Gaetano Donizetti. 44 illustrations. 186pp.
22110-5 Pa. $2.00

ANTONIO STRADIVARI: HIS LIFE AND WORK, W. H. Hill, et al. Great work of musicology. Construction methods, woods, varnishes, known instruments, types of instruments, life, special features. Introduction by Sydney Beck. 98 illustrations, plus 4 color plates. 315pp. 20425-1 Pa. $3.00

MUSIC FOR THE PIANO, James Friskin, Irwin Freundlich. Both famous, little-known compositions; 1500 to 1950's. Listing, description, classification, technical aspects for student, teacher, performer. Indispensable for enlarging repertory. 448pp.
22918-1 Pa. $4.00

PIANOS AND THEIR MAKERS, Alfred Dolge. Leading inventor offers full history of piano technology, earliest models to 1910. Types, makers, components, mechanisms, musical aspects. Very strong on offtrail models, inventions; also player pianos. 300 illustrations. 581pp. 22856-8 Pa. $5.00

KEYBOARD MUSIC, J.S. Bach. Bach-Gesellschaft edition. For harpsichord, piano, other keyboard instruments. English Suites, French Suites, Six Partitas, Goldberg Variations, Two-Part Inventions, Three-Part Sinfonias. 312pp. 8⅛ x 11.
22360-4 Pa. $5.00

COMPLETE STRING QUARTETS, Ludwig van Beethoven. Breitkopf and Härtel edition. 6 quartets of Opus 18; 3 quartets of Opus 59; Opera 74, 95, 127, 130, 131, 132, 135 and Grosse Fuge. Study score. 434pp. 9⅜ x 12¼. 22361-2 Pa. $7.95

COMPLETE PIANO SONATAS AND VARIATIONS FOR SOLO PIANO, Johannes Brahms. All sonatas, five variations on themes from Schumann, Paganini, Handel, etc. Vienna Gesellschaft der Musikfreunde edition. 178pp. 9 x 12. 22650-6 Pa. $4.00

PIANO MUSIC 1888-1905, Claude Debussy. Deux Arabesques, Suite Bergamesque, Masques, 1st series of Images, etc. 9 others, in corrected editions. 175pp. 9⅜ x 12¼. 22771-5 Pa. $4.00

INCIDENTS OF TRAVEL IN YUCATAN, John L. Stephens. Classic (1843) exploration of jungles of Yucatan, looking for evidences of Maya civilization. Travel adventures, Mexican and Indian culture, etc. Total of 669pp.
20926-1, 20927-X Pa., Two vol. set $5.50

LIVING MY LIFE, Emma Goldman. Candid, no holds barred account by foremost American anarchist: her own life, anarchist movement, famous contemporaries, ideas and their impact. Struggles and confrontations in America, plus deportation to U.S.S.R. Shocking inside account of persecution of anarchists under Lenin. 13 plates. Total of 944pp.
22543-7, 22544-5 Pa., Two vol. set $9.00

AMERICAN INDIANS, George Catlin. Classic account of life among Plains Indians: ceremonies, hunt, warfare, etc. Dover edition reproduces for first time all original paintings. 312 plates. 572pp. of text. 6⅛ x 9¼.
22118-0, 22119-9 Pa., Two vol. set $8.00
22140-7, 22144-X Clothbd., Two vol. set $16.00

THE INDIANS' BOOK, Natalie Curtis. Lore, music, narratives, drawings by Indians, collected from cultures of U.S.A. 149 songs in full notation. 45 illustrations. 583pp. 6⅝ x 9⅜.
21939-9 Pa. $5.00

INDIAN BLANKETS AND THEIR MAKERS, George Wharton James. History, old style wool blankets, changes brought about by traders, symbolism of design and color, a Navajo weaver at work, outline blanket, Kachina blankets, more. Emphasis on Navajo. 130 illustrations, 32 in color. 230pp. 6⅛ x 9¼.
22996-3 Pa. $5.00
23068-6 Clothbd. $10.00

AN INTRODUCTION TO THE STUDY OF THE MAYA HIEROGLYPHS, Sylvanus Griswold Morley. Classic study by one of the truly great figures in hieroglyph research. Still the best introduction for the student for reading Maya hieroglyphs. New introduction by J. Eric S. Thompson. 117 illustrations. 284pp.
23108-9 Pa. $4.00

THE ANALECTS OF CONFUCIUS, THE GREAT LEARNING, DOCTRINE OF THE MEAN, Confucius. Edited by James Legge. Full Chinese text, standard English translation on same page, Chinese commentators, editor's annotations; dictionary of characters at rear, plus grammatical comment. Finest edition anywhere of one of world's greatest thinkers. 503pp.
22746-4 Pa. $4.50

THE I CHING (THE BOOK OF CHANGES), translated by James Legge. Complete translation of basic text plus appendices by Confucius, and Chinese commentary of most penetrating divination manual ever prepared. Indispensable to study of early Oriental civilizations, to modern inquiring reader. 448pp.
21062-6 Pa. $3.50

THE EGYPTIAN BOOK OF THE DEAD, E.A. Wallis Budge. Complete reproduction of Ani's papyrus, finest ever found. Full hieroglyphic text, interlinear transliteration, word for word translation, smooth translation. Basic work, for Egyptology, for modern study of psychic matters. Total of 533pp. 6½ x 9¼.
EBE 21866-X Pa. $4.95

BUILD YOUR OWN LOW-COST HOME, L.O. Anderson, H.F. Zornig. U.S. Dept. of Agriculture sets of plans, full, detailed, for 11 houses: A-Frame, circular, conventional. Also construction manual. Save hundreds of dollars. 204pp. 11 x 16.
21525-3 Pa, $5.95

HOW TO BUILD A WOOD-FRAME HOUSE, L.O. Anderson. Comprehensive, easy to follow U.S. Government manual: placement, foundations, framing, sheathing, roof, insulation, plaster, finishing — almost everything else. 179 illustrations. 223pp. 7⅞ x 10¾.
22954-8 Pa. $3.50

CONCRETE, MASONRY AND BRICKWORK, U.S. Department of the Army. Practical handbook for the home owner and small builder, manual contains basic principles, techniques, and important background information on construction with concrete, concrete blocks, and brick. 177 figures, 37 tables. 200pp. 6½ x 9¼.
23203-4 Pa. $4.00

THE STANDARD BOOK OF QUILT MAKING AND COLLECTING, Marguerite Ickis. Full information, full-sized patterns for making 46 traditional quilts, also 150 other patterns. Quilted cloths, lamé, satin quilts, etc. 483 illustrations. 273pp. 6⅞ x 9⅝.
20582-7 Pa. $3.50

101 PATCHWORK PATTERNS, Ruby S. McKim. 101 beautiful, immediately useable patterns, full-size, modern and traditional. Also general information, estimating, quilt lore. 124pp. 7⅞ x 10¾.
20773-0 Pa. $2.50

KNIT YOUR OWN NORWEGIAN SWEATERS, Dale Yarn Company. Complete instructions for 50 authentic sweaters, hats, mittens, gloves, caps, etc. Thoroughly modern designs that command high prices in stores. 24 patterns, 24 color photographs. Nearly 100 charts and other illustrations. 58pp. 8⅜ x 11¼.
23031-7 Pa. $2.50

IRON-ON TRANSFER PATTERNS FOR CREWEL AND EMBROIDERY FROM EARLY AMERICAN SOURCES, edited by Rita Weiss. 75 designs, borders, alphabets, from traditional American sources printed on translucent paper in transfer ink. Reuseable. Instructions. Test patterns. 24pp. 8¼ x 11.
23162-3 Pa. $1.50

AMERICAN INDIAN NEEDLEPOINT DESIGNS FOR PILLOWS, BELTS, HANDBAGS AND OTHER PROJECTS, Roslyn Epstein. 37 authentic American Indian designs adapted for modern needlepoint projects. Grid backing makes designs easily transferable to canvas. 48pp. 8¼ x 11.
22973-4 Pa. $1.50

CHARTED FOLK DESIGNS FOR CROSS-STITCH EMBROIDERY, Maria Foris & Andreas Foris. 278 charted folk designs, most in 2 colors, from Danube region: florals, fantastic beasts, geometrics, traditional symbols, more. Border and central patterns. 77pp. 8¼ x 11.
USO 23191-7 Pa. $2.00

Prices subject to change without notice.
Available at your book dealer or write for free catalogue to Dept. GI, Dover Publications, Inc., 180 Varick St., N.Y., N.Y. 10014. Dover publishes more than 150 books each year on science, elementary and advanced mathematics, biology, music, art, literary history, social sciences and other areas.